The Greenwich Guide to Stars, Galaxies and Nebulae

The complete introduction to astronomy beyond the
Solar System
* The birth of the Universe
* The ageing Sun — an ordinary star
* Life and death among the stars
* Searching beyond the Galaxy — Catherine wheels
  in the sky
* The mysteries of deep space — black holes and
  quasars
* Beyond the frontiers . . .

Stuart Malin is the former Head of the Astronomy
Department at the Old Royal Observatory, Greenwich.

Cambridge University Press

ISBN 0-521-37777-3

9 780521 377775

The Greenwich Guides to Astronomy

The *Greenwich Guides* are a series of books on astronomy for the beginner. Each volume stands on its own but together they provide a complete introduction to the night sky, everything it contains, and how astronomers are discovering its secrets. Written by experts from the Old Royal Observatory at Greenwich, they are right up to date with the latest information from space exploration and research and are suitable for observers in both the northern and southern hemisphere.

Available now
The Greenwich Guide to Stargazing
The Greenwich Guide to the Planets
The Greenwich Guide to Stars, Galaxies and Nebulae
The Greenwich Guide to Astronomy in Action

The Old Royal Observatory, Greenwich, London is open daily to visitors. It is the home of Greenwich Mean Time and of the Greenwich Meridian which divides east from west. It also houses the largest refracting telescope in Great Britain.
If you would like more information, please write to: The Marketing Department, National Maritime Museum, Greenwich, London SE10 9NF.

The Greenwich Guide to
Stars, Galaxies and Nebulae

Stuart Malin

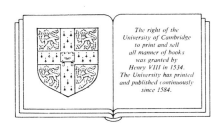

The right of the
University of Cambridge
to print and sell
all manner of books
was granted by
Henry VIII in 1534.
The University has printed
and published continuously
since 1584.

CAMBRIDGE UNIVERSITY PRESS
Cambridge
New York   New Rochelle   Melbourne   Sydney

First published by George Philip Limited,
59 Grosvenor Street, London W1X 9DA

This edition published by the Press Syndicate of
the University of Cambridge
32 East 57th Street, New York, NY 10022, USA

First published in this edition 1989

British Library Cataloguing in Publication Data

Malin, S. R. C. (Stuart Robert Charles), *1936–*
   The Greenwich guide to the stars,
   galaxies and nebulae.
   1. Astronomy
   I. Title
   520
ISBN 0-521-377773

Library of Congress CIP data available

© The Trustees of the National Maritime Museum 1989

Printed in Hong Kong

Acknowledgements

I am grateful to many people for their assistance in the preparation of this book. In particular I would like to thank Rosaly Lopes, David Hughes and Harry Ford for suggestions on the text, Carole Stott, Lindsey Macfarlane and Victoria Salter for help in assembling the illustrations, Marilla Fletcher who did much more than the typing and Lydia Greeves of George Philip who nursed it into print. While the illustrations came from many sources, as acknowledged below, special mention should be made of the photography of David Malin and the draughtsmanship of Paul Doherty.

I am grateful to the following for permission to reproduce their illustrations: Anglo-Australian Telescope Board: pp. 11, 43, 44, 51, 53, 54, 56, 57, 58, 61, 64, 66, 68, 71, 73, 74, 76, 79, 83; Department of Energy: p. 19; Paul Doherty: pp. 31, 41, 77, 80; European Southern Observatory: pp. 47, 72; NASA: p. 20; National Maritime Museum: pp. 17, 28, 30; National Optical Astronomy Observatories: pp. 10, 15, 22, 25, 27, 35, 45, 50, 52, 60, 65, 70, 84, 85, 86; Royal Observatory, Edinburgh: pp. 2, 8; Whitworth Art Gallery, University of Manchester: p. 12. All artwork is by Paul Doherty.
Jacket illustrations: Anglo-Australian Telescope Board (front); Jim Stevenson (back).

FRONTISPIECE *The Lagoon Nebula, M8. This huge cloud of glowing gas also contains smaller dark clouds of light-absorbing dust, known as 'Bok globules'. Although just visible to the naked eye, a long-exposure photograph is required to show the Lagoon Nebula's true beauty.*

JACKET ILLUSTRATIONS: *The spiral galaxy NGC 2997 (front) and the Old Royal Observatory, Greenwich, London (back).*

Contents

Introduction

What could be more restful than gazing up at the beauty of a starlit sky, the familiar pattern of the constellations apparently fixed for eternity? But this sense of peaceful tranquillity is an illusion. Each spot of light is a blazing, self-destructive inferno, some stars sinking quietly into oblivion and others flaring up in a spectacular final explosion. And deep in the heart of turbulent gas clouds new stars are being born all the time. Beyond the stars are the hazy, indistinct outlines of galaxies, each home to tens of thousands of millions of stars.

This book reveals how astronomers have used starlight to build up a picture of the life and death of stars, and how observing far distant galaxies has helped us understand our own and the origin of the Universe itself. Without the strange and fascinating processes going on in the depths of space, neither the Earth nor Mankind would have come into existence.

1 · *The Empty Universe*

There are few more inspiring sights than the night sky on a clear, moonless night well away from city lights. Thousands of points of light shine down on us just as they did on our remote ancestors, and as they will continue to shine for millions of years, regardless of what we do on Earth. But just how permanent are they? Where did they come from and how will they end? These and many other questions spring to mind, and it is the purpose of this book to try and answer some of them. Not all the answers are known, of course, and some of the most interesting questions (those that start with 'why') are beyond the scope of science. Nevertheless, a great deal *is* known and it makes a fascinating story.

The most important point about the Universe is that it is virtually empty. This is not obvious from where we sit, on a comfortable little planet close to a nice warm Sun, so let us suppose that you have been removed and put down at random anywhere in the Universe. There is a chance that you will find yourself close to a star, but this possibility is so unlikely that it can be ignored. So what will it be like? On average, the nearest star will be 30,000,000,000,000,000,000 kilometres away and so faint that you could not see it even with the most powerful telescope. And between you and the nearest star: nothing. This means that it will be very dark – darker than the blackest night because there will not even be starlight—and bitterly cold. The temperature will be only a few degrees above absolute zero: about 270°C below freezing.*

Dark though it is, the sky will not be totally black. Through a powerful telescope you may be able to see a few very faint and slightly fuzzy spots of light. These are not individual stars but galaxies, consisting of tens of thousands of millions of stars whose combined light you can just detect. The Universe consists of galaxies and empty space, with very much more of the latter. Less than 0.000002 per cent of space is occupied by galaxies, so you would not be far wrong if you said it was completely empty.

Let us now leave this bleak, cold, random place in the Universe and return home to Earth. It is now clear why our night sky is so much richer than just described—we actually live inside a galaxy. Just about everything you can see in the sky with the naked eye is within our own Galaxy (which is always spelt with a capital G): the only exceptions are the few neighbouring galaxies that make up our local group of galaxies. To see any of the other galaxies in the Universe you would need a good telescope and even then you would see only a faint blur. Nevertheless, these distant galaxies are of great importance because they are the building blocks of the Universe, and it is only by studying them that we can learn anything about the Universe as a whole.

The first important question is their distribution in space. It is easy enough to map the positions of bodies on the celestial sphere by

---

*All matter is made up of minute particles which are constantly in motion. Temperature is a measure of this motion. When all the particles are stationary the temperature is absolute zero—minus 273°C—and it cannot decrease below that value. Scientists measure temperature on the absolute, or Kelvin scale, with 0K for absolute zero, 273K for the freezing point of water and 373K for its boiling point.

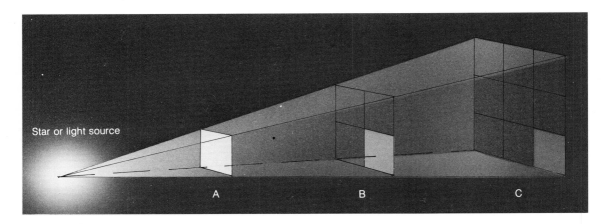

Star or light source

A   B   C

ABOVE *An object twice as far away as another will appear to be only a quarter as bright. The starlight that illuminates area A has to illuminate four times that area at twice the distance (B), so a star seen from B would appear only a quarter as bright as one seen from A, or one-ninth as bright from three times the distance (C). This is an example of the inverse square law.*

measuring angles (see *The Greenwich Guide to Stargazing*), but to get the full picture we also need to know how far away they are. To get their relative distances, we can make the simple assumption that all galaxies give out the same amount of light, so the reason that some appear brighter and others fainter is because the fainter ones are further away. This assumption is not always reliable when applied to individual galaxies, but holds fairly well when considering the average of large numbers. If one galaxy is twice as far away as another, it will appear to be

Cluster of galaxies in the constellation of Fornax, their amoeba-like shapes contrasting with the point images of the foreground stars belonging to our own Galaxy. The brighter nearby stars show as crosses, but the spikes are produced by the telescope and are not real.

only a quarter as bright. (You can see this with a row of street lights—they are all the same brightness, but the distant ones appear fainter.) What we find is that, apart from a tendency to occur in groups, galaxies are more or less uniformly distributed throughout space. There is no obvious 'centre of the Universe' and no direction (except where the view is obscured by material in our own Galaxy) where galaxies cannot be seen.

Another characteristic that can be measured is their line-of-sight velocity—how fast they are approaching us or receding from us. This *radial velocity* is calculated from the *doppler effect*. We have all noticed the doppler effect in sound: when a car drives past you sounding its horn, the pitch of the note is higher when it is approaching than when it has passed. If you know the true pitch of the car's horn, you can use the speed of sound and the change of pitch to deduce the car's speed. Similarly with the light from galaxies: if a galaxy is approaching, the 'pitch' or frequency of its light is raised and becomes bluer; if it is receding the light is shifted towards the red end of the spectrum. What we find is that just about all the galaxies are red-shifted—they are moving away from us. What is more, the red-shift is proportional to the distance; the further a

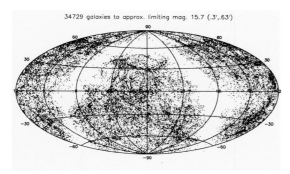

34729 galaxies to approx. limiting mag. 15.7 (.3',.63')

The distribution of galaxies over the sky as a whole, drawn so that equal areas of the map represent equal areas of the sky. Except where the view is obscured by our own Galaxy, the distribution can be seen to be more or less uniform.

galaxy is away from us, the faster it is receding. This gives us the important observational result that the Universe is expanding!

This has significant consequences. The faintest (and hence most distant) galaxies that can be observed are receding from us at speeds that are an appreciable proportion of the speed of light. At some still greater distance it is reasonable to assume that galaxies are receding faster than the speed of light, but we can never see them, because we are receding from them at the same speed, so their light will never overtake us. In fact, we can never find out anything about them, because it is impossible to transmit information of any kind faster than the speed of light. This means that there is a limit to the size of the observable Universe.

You may think that because everything is receding from us uniformly in all directions we must be at the centre of the Universe, but this is not so. Imagine an observer on a distant, fast-receding galaxy. He will say that he is at rest, and that we are receding from him. He will also see all the other galaxies receding, the nearby ones slowly and the distant ones rapidly, just as we do. The expanding Universe looks just the same from

wherever you view it, like an infinite pudding full of raisins (clusters of galaxies), as Sir Fred Hoyle has described it. As the pudding cooks it expands and all the raisins move away from one another (though the individual raisins do not expand); the rate at which any pair of raisins separate is directly related to their distance apart. Again, the view is the same from any raisin and none can claim to be at the centre of the pudding— remember it is an infinite pudding, so no boundaries can be seen.

Now let us have a look at what things were like in the past. If the galaxies are getting further apart at present, then they must previously have been closer together. As we go further and further back in time they get closer and closer together until we reach a time when everything in the Universe was jam-packed together in one place as a superdense blob. This is how the Universe started about 15 thousand million years ago.

On a timescale of thousands of millions of years, one second might seem totally un-important, but it is what happened within the first second of the existence of the Universe that determined how it would evolve into what we see today. It is tempting to ask what was there before the Universe came into existence, but this is outside the realm of scientific speculation for the same reason as the Universe beyond the observable limit—there is no possible way of obtaining information about it. We are also in trouble with the first few millionths of a second, but for a different reason. We do not know enough about the behaviour of matter at the unimaginably high temperatures and densities that existed then. All we can say is that the Universe suddenly came into existence all at one place with an almighty bang, incomparably more powerful than anything that has ever happened

The spiral galaxy NGC 253, seen in an edge-on view. A typical galaxy of this sort contains thousands of millions of individual stars.

12

since, which flung the material far out into space and was directly responsible for the expansion of the Universe which is still going on.

Material as we know it did not exist then. There were no atoms, with a dense central nucleus of protons and neutrons surrounded by a cloud of electrons, because these subatomic particles had not yet formed. There were not even the still more fundamental particles, like quarks and mesons. By the time the Universe was one ten-thousandth of a second old, quarks and mesons had been forged in the inferno, the temperature had dropped to a million million K and the density was down to 10 million million times that of lead—the sort of density the Great Pyramid would have if the entire Earth were compressed into it. As the Big Bang continued, the Universe spread out and its temperature and density dropped, allowing subatomic particles to form. By the time 10 seconds had elapsed, many subatomic particles were welded together to form the nuclei of atoms, and radiation had taken over from atomic interactions as the main driving force. Radiation—energy in the form of heat and light—continued to dominate for the next million years, by which time the temperature had dropped to below 10,000K and the Universe was so spread out that a volume the size of the Earth would contain only one tonne of material. Under these conditions, electrons would be moving slowly enough to be captured by the atomic nuclei to make real atoms, and the process of star formation could begin. We will consider this process later, but it is important to realize that the Universe was a million years old before the first stars were formed. This means we can find out nothing about that first million years by observing even the oldest stars, but have to rely instead on our knowledge of the laws of physics.

There is, however, one important clue to the early state of the Universe and this comes from

God creating the Universe, as depicted by William Blake in 1824.

the radiation released in the initial explosion. As time went by, this gradually became less intense. It is still there, though now it is so weak that it corresponds to the 'glow' that would come from material just 3° above the absolute zero of temperature. Nevertheless, two American scientists, Arno Penzias and Robert Wilson, were able to detect this faint background radiation and thus confirm the Big Bang theory of the origin of the Universe. For this work they were awarded the Nobel Prize for physics in 1978.

After the first million years had elapsed, the Universe was a huge cloud of uniform, thin gas rather hotter than the surface of the Sun. Some time between then and the present day it changed dramatically, so that it is now a large number of small, fairly dense blobs (the galaxies) with nothing in between. This change has been brought about mainly through the action of gravity. When the cloud was very hot, the atoms of which it was composed would bump into one another and bounce off, thus keeping the cloud well stirred, but as the temperature and density dropped, the collisions became less frequent and less violent so that the stirring mechanism became less effective.

At this stage, the much weaker force of gravity became more important than radiation. Every particle in the Universe attracts every other particle with a force that depends on the masses of the particles and their distance apart. The closer they are together, the greater the gravitational force trying to draw them still nearer to each other, so if a group of particles clusters by chance, gravity will tend to bind them closer still. This gathering effect is opposed by the thermal stirring, trying to bounce the particles away from one another, but as the temperature drops and the density increases, the influence of gravity takes over. Although the gas cloud was uniform, it was not perfectly uniform, so there were many regions where the density was a little above the average. Those regions tended to condense, leaving bigger gaps of lower density

between them. By this process, over many millions of years, the Universe became divided up into innumerable blobs of gas which would eventually become clusters of galaxies. At the same time, the Universe continued to expand, increasing the distance between the blobs.

Within the blobs, the same process was going on on a smaller scale, with each one breaking up into millions of much smaller blobs which would form into stars. When a galaxy was still largely composed of gas, collisions between particles would be quite frequent, and this would help the process of knocking the galaxy into its final shape. Once the gas had condensed into stars, there would be fewer collisions (though still some gravitational energy-exchange as stars passed close to one another) and the evolutionary process would slow down. It is probably the presence or absence of gas clouds in galaxies after stars have formed as well as their rate of rotation that accounts for the diversity in their structure (see Chapter 5), but the processes of galaxy formation are not yet well understood.

So the Universe started with a Big Bang; how will it end? There are two distinct possibilities. One is that the expansion will go on for ever, with the groups of galaxies getting further and further apart and the average density getting nearer and nearer to zero. The other possibility is that the gravitational attraction of the groups of galaxies for one another will eventually overcome the expansion and start to draw them together again. Once this started to happen, the contraction would accelerate until all the galaxies came crashing together in something like the Big Bang in reverse. Which of these two possibilities will occur depends on the total amount of material in the Universe and at present this quantity is not well enough known for a prediction to be made. It is similar to the launching of a spacecraft: if it is launched at less than a certain speed the gravitational attraction of the Earth will prove too strong and it will eventually fall back to Earth. But if it is launched at a speed great enough to counteract the pull of the Earth, it will escape forever. Even if it is the ultimate fate of the Universe to collapse in on itself, we can take comfort from the fact that it is still expanding at present so we are not yet halfway through the process even after 15 thousand million years!

These two galaxies, NGC 4567 and NGC 4568 in the constellation of Virgo, appear to be linked by a bridge of hydrogen.

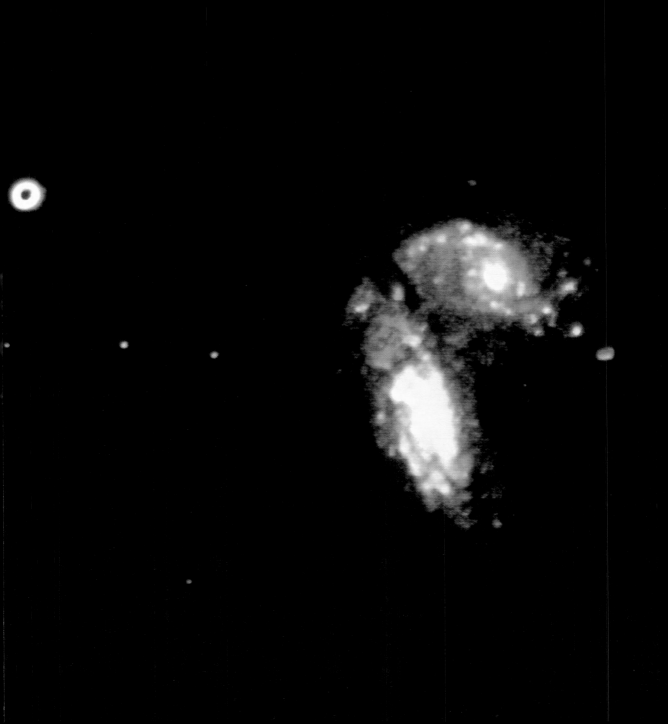

2 · An Ordinary Star

What special equipment do you need to see a star during the daytime? The answer is simply a pair of eyes; indeed it would be dangerous to use anything more powerful because the star is, of course, the Sun. It is very easy to scorch a piece of wood by focusing the Sun on it with a small magnifying glass, and this is exactly what you would be doing to your eyes if you were to look at the Sun through binoculars. So never do it. If you want to see a magnified image of the Sun, it is best to project the light through binoculars or a telescope on to a white card, as described in *The Greenwich Guide to Stargazing*. Even looking directly at the Sun with the naked eye can be harmful, unless you use a filter.

The Sun is a particularly convenient star to start with, not only because we have a grandstand view of it (the next nearest star is more than a quarter of a million times as far away as the Sun), but also because, apart from being a little more massive than most, it is just about as typical an example of a star as could be found anywhere. Many of the Sun's features, such as sunspots and flares, have not been seen on any other star, but that is because the other stars are too far away for such detail to be seen; it is likely that most stars have the same characteristics. This may even apply to the system of planets that goes around the Sun. Although the evidence is very slight at present, it may be that planetary systems are very common throughout the Universe. It is not impossible that, somewhere in space, a bug-eyed monster is even now speculating on whether we exist!

The early history of the Sun and planets and how they condensed out of a gas cloud some 4600 million years ago is told in *The Greenwich Guide to the Planets*, but there we were more concerned with the planets and here we are looking at the Sun. The first important point is that it started life about 5 thousand million years ago, so it is a relatively young star, less than half the age of the Universe. Clearly the Sun is not one of the stars that first started to condense when the Universe was still young, and this makes an important difference to its composition. The very first stars were composed of only the lightest of chemical elements: mostly hydrogen, some helium and traces of lithium and beryllium. These were the elements that had been formed in the Big Bang and were present when the primeval gas cloud first broke up into galaxies. The heavier elements familiar to us on Earth, carbon, oxygen, nitrogen, iron etc, have been produced by nuclear reactions inside stars. The gas cloud from which our Sun formed had been enriched with these heavier elements, formed in an earlier generation of stars and then dispersed through the Galaxy when these stars exploded.

Astronomers distinguish between the earlier types of star and the more recent, metal-rich stars by calling them Population II and Population I respectively. (To an astronomer, anything other than hydrogen or helium counts as a metal.) Although the Sun is metal-rich in comparison with Population II stars, the heavier elements account for less than 2 per cent of the mass of its

The setting Sun is reddened and distorted by the Earth's atmosphere.

surface, with helium providing 25 per cent and hydrogen, the lightest of all elements, making up the remaining 73 per cent. Even so, the heavier elements have an important part to play in the processes occurring within the Sun. And, more importantly for us, without heavier elements there could be no Earth.

The Sun started as a large, tenuous ball of gas which contracted under the influence of its own gravitational attraction. If gravity had been the only force at work, the gas would very rapidly have collapsed to form a tiny ball, but there were other forces working against gravity to keep the gas spread out. One of these forces is rotation: a gently rotating ball of gas will spin faster and faster as it contracts until the point is reached

where it will break up unless it can lose some of its *angular momentum*, as the spin energy is called. In the case of the Sun, much of the angular momentum was left behind in the orbital motions of the planets, and more was lost by ejecting hydrogen in the form of a wind that blew out from the Sun through the Solar System. The end result is that the Sun's rotation is now a sedate one revolution in 27 days.

The other, and more important, force that keeps the Sun inflated is gas pressure. The atoms in a gas are all moving about, bouncing off one another and bumping into anything that gets in their way. Each individual bump is quite trifling, but the total effect can be a very substantial pressure. There are two ways of increasing gas pressure: one is to increase the density of the atoms—double the number of atoms in a given volume and you will double the pressure. The other way is to increase the average speed of the atoms by heating them up. The temperature of a gas is simply a measure of the average speed of the atoms of which it is composed. Anyone who has experimented with a bicycle pump will be familiar with these facts. If you put your finger over the hole, it is easy to push the pump handle down a little way, but it gets more and more difficult as you go further. This is because you are increasing the density of the gas in the pump by compressing it into a smaller volume and, as you do so, it presses back at you. You will also notice that when you pump up a tyre the bottom end of the pump gets hot. As you compress the gas, you also heat it up.

The same things happen to our pre-Sun gas ball: as it is compressed by gravity, the pressure and the temperature both go up. At some stage, when it is small enough, the Sun will be glowing white hot, the gas pressure outwards will be high enough to balance the gravitational pressure inwards and the ball will stop shrinking. This state of affairs could not last, however, as the Sun is radiating its heat away into surrounding space. The tiny proportion of this radiation that is intercepted by the Earth prevents our planet from freezing up and enables life to exist. If no other processes were going on, the Sun would cool, the gas pressure would drop and the ball would get smaller and smaller until eventually it would end up as a small, solid, cold lump. This would happen very quickly in astronomical terms, in just a few million years, and the Solar System would long ago have become cold and dead. Obviously the Sun must have some other source of heat besides that provided by gravitational compression.

At one time it was thought that the Sun was burning like a fire on Earth, combining its inflammable constituents with oxygen to give out heat and light. But even a bonfire the size of the Sun would burn itself out in just a few thousand years if it was producing heat and light at the rate the Sun does. In fact, the source of heat in the Sun is the same as that in a hydrogen bomb: a thermonuclear reaction in which hydrogen is converted into helium with the release of large quantities of energy. Such a reaction can take place only at a temperature of several million degrees, so it was not until the Sun had been heated to such a temperature by gravitational compression that the thermonuclear reaction could start. Once under way this reaction could produce as much heat as was required to replace that radiated away into space, and could carry on doing so for thousands of millions of years.

You cannot get something for nothing on the Sun any more than you can on Earth, so where does all this energy come from? It is the result of converting mass into energy according to Einstein's famous equation $E = mc^2$, where E denotes energy, m mass and c the velocity of light. Since c is a really huge number, the destruction of a very small quantity of mass releases an enormous amount of energy. The conversion of just one gramme would provide enough energy to make a thousand million cups of tea. The nucleus of the helium atom that results from a

The giant mushroom cloud produced by a thermo-nuclear explosion on Earth. The same reactions are going on continuously in the Sun.

thermonuclear reaction is very slightly (0.7 per cent) less massive than the hydrogen nuclei from which it is formed by fusion. The extra mass is converted into energy in the form of heat and light. To produce the vast amount of energy that the Sun radiates, it has to convert 4 million tonnes of mass into energy every second. To do this, it has to convert nearly 600 million tonnes of hydrogen into helium every second. This may seem a quite staggering quantity until you realize that the Sun is so huge that it could carry on

doing this for 80 thousand million years before its hydrogen was all used up.

It is not by chance that the energy generated within the Sun is just the same as the energy it radiates into space. If more energy were produced in the Sun's interior, the temperature and hence the pressure would go up. This would cause the Sun to expand until its surface area was large enough to radiate away the extra energy, and balance would again be restored. Curiously, this new, enlarged Sun would have a cooler surface than at present, though its deep interior would be hotter. Similarly, if the rate of energy production went down, we would end up with a smaller Sun with a hotter surface. Some stars have difficulty

19

in getting the balance right and oscillate about the equilibrium position, changing from large, cool stars to smaller, hotter ones and back again in a matter of days. Fortunately for us, the Sun is more stable. If it were not, the Earth would regularly be scorched and frozen, making it quite unsuitable for life of any sort.

It is clear from this that the temperature of the Sun is not the same all the way through. The

visible surface of the Sun, or *photosphere* to use the technical term, has a temperature of about 6000K. This can be deduced from its colour in the same way that furnace temperatures are measured on Earth. As the temperature of a furnace goes up, its contents glow first red hot, then orange, yellow, white and, at the highest temperatures, blue hot. By comparing the colour with that of an electrically heated wire, the

Photographing the Sun in hydrogen-alpha light reveals much more detail than can be seen in white light.

temperature can be measured with some precision. Deeper down in the Sun, where it is insulated from the coldness of space by the overlying layers, the temperature is higher, reaching a searing 15 million K in the central regions where the thermonuclear reactions take place. The density of the Sun is not uniform either. Its average density is not much greater than that of water, but it varies from that of a thin gas at the surface to more than ten times the density of lead at its centre. This is because the compression caused by the weight of the overlying layers increases the nearer you go to the centre. With a density as high as that of lead, you might think that the heart of the Sun is very solid indeed, but remember that the temperature is also incredibly high—high enough to keep the Sun gaseous all the way through.

So that is what the Sun is like at present: a continuously exploding hydrogen bomb in the centre of a large ball of gas, mostly composed of hydrogen but with a substantial proportion of helium. It is hot and dense at the centre and much cooler and thinner at the surface. But will it always be like that? We have seen that it can continue in its present state for thousands of millions of years, but eventually most of the hydrogen in its core will have been converted into helium and as this central furnace runs out of fuel it will no longer be able to produce the energy required to resist the inward pressure from the rest of the Sun. As the core is compressed, the surrounding material (still mostly hydrogen) will get hot enough for a thermonuclear reaction, so there will be a 'dead' core surrounded by a burning shell of hydrogen. This shell will produce more heat than the original central core, and this will cause the Sun to inflate to about a hundred times its present diameter, swallowing up the planet Mercury on its way and making the Earth uninhabitable.

Although the total radiation from the Sun will be about a thousand times greater than at present, the surface will be only red hot—the Sun will have become a *red giant*.

In supporting such a huge star, the internal furnace uses up its nuclear fuel at a prodigious rate. Even though the internal temperature will get hot enough for nuclear reactions among the heavier elements, for example the fusion of helium to produce carbon and oxygen, it will not be long (in astronomical terms) before the furnace has used up all its fuel. As this happens, the Sun will shrink and its surface will get hotter. There is now nothing to replace the energy radiated away into space except what is generated by the Sun's collapse under its own gravitational pressure. It will become a tiny *white dwarf*, not much bigger than the Earth, but very, very dense. As the cooling continues, it will become a *black dwarf*, just a small, heavy, cold, dark lump floating through space until the end of time.

Of course, much of this is speculation, based on observations of other stars and on the laws of physics as we understand them at present. Our knowledge of the processes involved is far from complete, which is just as well. Astronomy would be a very dull subject if all the answers were already known. We can, however, be much more confident about the results obtained from direct observation of features on the Sun. Similar phenomena could probably be seen on most other stars, if they were not so far away.

The Sun's most obvious characteristic is that it is not the 'pure, unblemished orb' that our forefathers believed it to be, but frequently shows small, dark markings. The ancient Egyptians' god Ra has a spotty face! Sunspots have been studied since the time of Galileo, and can easily be seen when the Sun's image is projected on to a piece of card by a small telescope (*do not look directly through the telescope*). The Sun is seldom completely free of them, and often it is possible to see quite a number, either occurring singly or,

21

more frequently, in pairs or in groups. A sunspot can last from days to months before it fades out. If the position of a sunspot is observed on successive days, it will be seen to move across the Sun's disc from east to west. This is because the Sun is rotating, and a study of sunspot movements gives a way of measuring the rotation rate. The answer you get depends on the latitude. Near its equator the Sun rotates once every 27 days, but towards its poles it takes nearly 32 days. There is no problem here; the Sun is not a solid body so there is no reason why it should all rotate at the same rate.

The Sun in hydrogen-alpha light on 26 June 1978, showing plages (the brighter areas), filaments (dark markings on the disc) and prominences (around the outer edge, or limb).

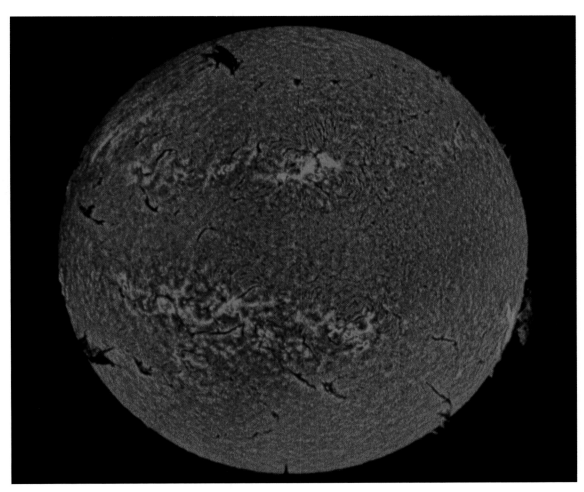

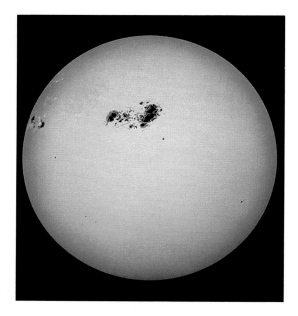

Longer-term studies of sunspots show that their numbers vary quite dramatically over a period of about 11 years. At sunspot maximum there may be as many as a hundred spots, but at minimum (1986 or 1987) the disc is sometimes completely free of them for months at a time. At the start of a sunspot cycle, the spots are most commonly seen at 30–40° above or below the Sun's equator, but as the cycle progresses they favour lower latitudes. This is clearly shown in the graphical presentation devised by the Greenwich astronomer Walter Maunder and known, for obvious reasons, as a butterfly diagram. Sunspots appear dark because they are cooler than their surroundings. They are also associated with localized regions of intense magnetism resulting from electric currents that are generated within the Sun. These currents are variable, though, and the direction of the magnetism in the spots of one cycle is opposite to that in the spots of the next cycle, so perhaps

ABOVE *The great sunspot of 1947 was one of the largest ever recorded. This artist's impression of its appearance on 7 April is based on photographs taken from the Mount Wilson Observatory in California, USA.*

BELOW *The butterfly diagram shows how the occurrence of sunspots varies with solar latitude and with time. A sunspot cycle lasts for about 11 years; at the start of each cycle, the spots tend to appear at higher latitudes, but become more common towards the Sun's equator as the cycle progresses.*

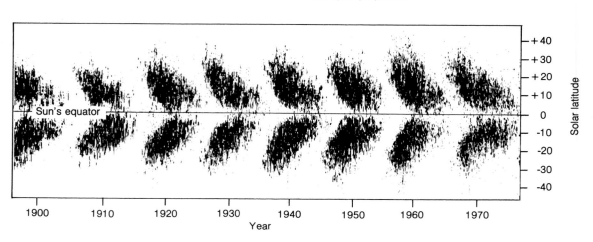

the true period of a sunspot cycle should be considered as 22 years rather than 11.

Sunspots are by no means the only features on the Sun, but to see the others special equipment is needed to cut down the intense background light. One means of doing this is to use a filter that transmits only the characteristic red light that is emitted by hydrogen. Through such a filter sunspots are seen to be surrounded by bright areas, and the whole disc is mottled. The mottling, called *supergranulation*, results from convection in the outer part of the Sun: hot regions welling up to the surface cool by radiating heat and light into space, spread out, and then sink down again. The brighter parts of the supergranulation are the hot gas that has just reached the surface and the darker parts are the cooler gas which is about to sink down. The whole surface is covered with these convection cells, hence the mottled appearance.

Under ideal observing conditions, when our atmosphere is particularly steady, even more detail can be seen. The supergranules are covered with a seething mass of tiny specks looking somewhat like sandpaper, except that the individual grains come and go over an interval of about 10 minutes. Although they appear tiny from Earth, the grains are in fact about 1000 kilometres across and result from turbulence in the Sun's atmosphere as heat is transmitted through it. This is called *granulation* and is produced by much the same process as super-granulation, though on a much smaller scale. *Spicules*, pointed structures thousands of kilo-metres high, can often be seen round the edges of the supergranules. When seen on the edge, or *limb*, of the Sun, they look like many blades of grass moving in the wind.

Spicules, granulation and supergranulation are always present on the Sun's surface, but

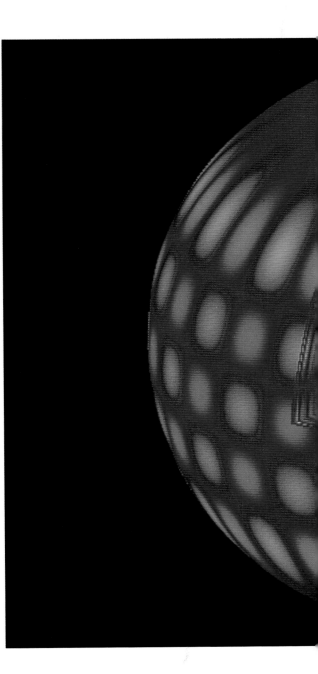

A computer simulation of how the outer regions of the Sun break up into a large number of convection cells, producing supergranulation on the surface.

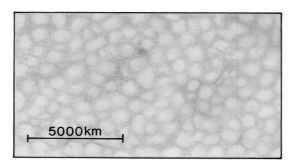

Artist's impression of solar granulation seen in hydrogen-alpha light. This fine-grained structure on the surface of the Sun is constantly changing and is difficult to see because of turbulence in the Earth's atmosphere.

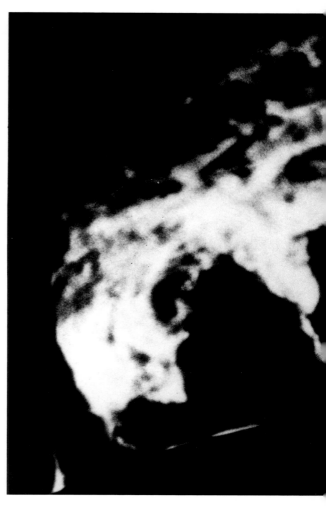

occasionally, particularly around the times of sunspot maximum, the Sun puts on a more spectacular display, with flares and prominences. The violent outbursts of energy known as *solar flares*, which last at most for an hour or so, are occasionally bright enough to show up on the Sun's disc even without a filter. Like sunspots, they appear to be associated with regions of intense magnetism. As well as light and heat, a flare radiates high-velocity streams of electrically-charged particles which cause dis-

Artist's impression of spicules on the surface of the Sun, with an image of the Earth included to give the scale.

plays of aurorae (the northern and southern lights), agitation of compass needles and radio interference when they encounter the Earth's atmosphere about a day after they are ejected. *Prominences* are also visible in hydrogen light on the Sun's disc, but they are best seen in profile on the limb. They are huge eruptions from the surface of the Sun which blast material out to

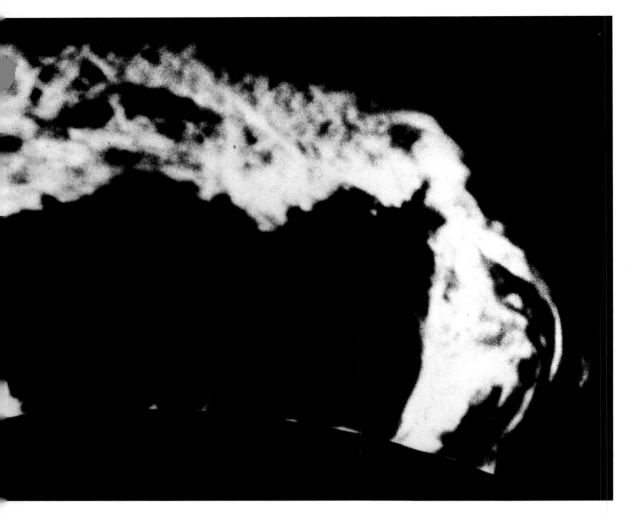

A huge prominence erupting from the surface of the Sun. This type of feature can be seen only when the much brighter disc of the Sun is obscured either by an eclipse or by a coronascope.

altitudes of tens of thousands of kilometres, or even occasionally shoot it out so far that it escapes into space. Interestingly, when the material falls back to the surface, it does not fall straight down as it would if gravity were the only force acting on it, but streams down along loops and arches which are the characteristic shapes of magnetic lines of force.

The outermost feature we can see from Earth is the Sun's crown, or *corona*, to use the Latin word. This occurs high above the surface but is so faint

27

that it can be seen only when the Sun's disc is completely obscured, for example during a total solar eclipse. But such eclipses last at most for a few minutes, are very infrequent and can be seen from only a limited region of the Earth's surface under cloudless conditions, so it is fortunate the

The solar corona, the tenuous but very hot outer atmosphere of the Sun, seen during a solar eclipse. The Sun's magnetic field produces the lineations in much the same way as a magnet lines up iron filings.

French astronomer, Bernard Lyot, invented what is known as the *coronagraph* in 1930. This reproduces the effect of an eclipse inside a telescope and permits the corona to be observed on any clear day from a high-altitude observatory. Even better results are obtained from a coronagraph on a satellite, outside the Earth's atmosphere. The corona is a tenuous gas that extends for over a million kilometres (several solar radii) above the Sun's surface. Once again, the structure of its brighter regions shows the

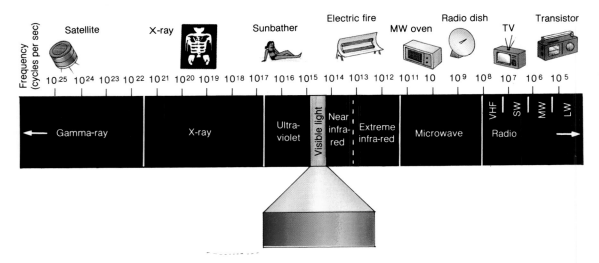

The electromagnetic spectrum. Stars emit radiation at all frequencies from radio waves to gamma rays.

characteristic shapes associated with magnetic lines of force. Remarkably, the corona has a temperature of well over 1,000,000K, but even so it is so tenuous that it emits only about a millionth of the total light from the Sun—less than that from a full Moon.

All the features described above can be seen in visible light, the sort to which our eyes are sensitive, but the Sun also emits radiation which the eye cannot detect. This ranges from beyond the red end of the spectrum—infra-red radiation and radio waves (the Sun was the first radio star to be detected)—to beyond the violet end—ultra-violet light and X-rays (see *The Greenwich Guide to Astronomy in Action*). The Earth's atmosphere absorbs most of the ultra-violet (though enough gets through to give you a suntan) and just about all the X-rays, so these can only be observed from a satellite. Much of the information obtained at these wavelengths confirms and extends what was found in visible light.

3 · Holes in the Sky

We now know that stars are not pinpricks in the fabric of the sky which allow the heavenly light to shine through, but it is easy to see why our ancestors thought of them that way. To the naked eye, the only obvious differences between stars are that some are brighter than others and that, at least among the brighter ones, there appears to be some variety of colour. But much more can be learned about stars than that. Much of the fascination of astronomy comes from the challenge of finding out as much as possible about a star from the faint glimmer of light that is the only evidence we have of its existence. It is remarkable how much can be discovered by combining our knowledge of the laws of physics with information gathered from starlight, but it has only proved possible as a result of years—in some cases centuries—of careful observation.

Some observations give us direct information about the star itself. For example, we would expect the colour of a star to appear the same from wherever in the Universe we were to observe it, unless there happened to be a cloud of dust in the way. But many observations, such as brightness, depend as much on our viewpoint as on the star: if a star is twice as far away as another star of a similar kind, it would appear to be only a quarter as bright. However, if we can measure the distance as well as the brightness of a star, we can deduce how luminous it really is.

Essentially, there are just three things you can do with the light from a star: you can measure the direction it is coming from, you can measure its intensity and you can split it up into its component colours. But this is like describing writing as making marks on a piece of paper. It is true as far as it goes, but gives no indication that, when expertly done, these measurements can convey a wealth of information.

The direction of the star tells us where it is in the sky, which is useful if you want to know where to point your telescope, but tells us very little about the star itself. Of greater interest are the tiny changes in position revealed by a sustained series of accurate measurements. If the position of a star is plotted relative to the background stars, and if the star is near enough, it will show a minute oscillation over a period of a year. This is known as *parallax* and arises because

A woodcut showing a philosopher looking through the fabric of the heavens to see what lies beyond. Despite its apparent age, it is now believed to have been drawn in the nineteenth century by the French astronomer Camille Flammarion.

The constellation Leo really does look like a lion, with his tail and hindquarters represented by three stars on the left, and a sickle of stars to the right representing his mane and chest. The handle of the sickle is the bright star Regulus.

we get a slightly different view of the stars as the Earth moves from one end of its orbit to the other. You can see the effect by holding up a finger at arm's length and looking at it first with the left eye and then with the right. It will appear to move relative to the far side of the room. By measuring the amplitude of the parallax, we can deduce the distance of the star (see *The Greenwich Guide to Stargazing*). Unfortunately, only a few hundred stars are close enough to have a measurable parallax, so the distances of the rest have to be deduced by a less direct method. Although very difficult, the measurement of distances *via* parallax has proved a vital first step in determining the more fundamental properties of a star.

When viewed through a telescope, it is found that very many stars are not isolated, like our Sun, but occur in pairs, known as *binaries*. Providing a pair is a true binary and not just a chance alignment of two stars at different distances, many years of measuring the position of one star relative to the other will show that one appears to rotate round its companion in an ellipse. This is unlikely to be the true orbit of the star, because we are probably seeing it at an

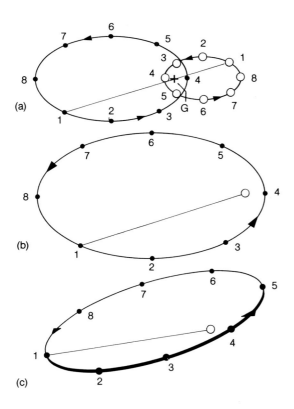

(a)

(b)

(c)

The orbit of a double star. The top diagram (a) shows how the two components of a binary star will each follow an elliptical orbit about the common centre of gravity (G). G is always on a line joining the stars, its exact position depending on the mass of each. For example, if the smaller star is half as massive as the larger one, G will divide the line in the ratio 2:1. (b) The orbit of one star relative to the other is also an ellipse, but we are rarely square-on to the orbit and usually get the distorted view shown in (c).

angle rather than head-on, but, by applying Kepler's laws (see *The Greenwich Guide to the Planets*), we can deduce the shape of the true orbit and its period. If we also know the distance of the star, we can convert the angular measurements into kilometres and obtain the true diameter of the orbit.

All this is not just for fun, but is the most direct means we have of obtaining the most fundamental of all a star's properties—its mass. It is explained in *The Greenwich Guide to the Planets* how the mass of the Sun can be found from the period and diameter of the orbit of one of its planets (or the mass of a planet from the orbit of one of its satellites) and just the same method will give us the mass of a double star from its period and orbital diameter. Unfortunately, there is a catch. What we end up with is the combined mass of both the stars that constitute the binary. This was also true when we measured the mass of the Sun, but the planets are so much smaller than the Sun that their masses can be ignored; this is not true for a binary star, where both components may have a similar mass.

There is a solution to this problem and again it comes from measurements of position, but this time relative to a fixed reference frame. Over a long enough period, most stars are found to change their positions slightly in the sky, each one proceeding steadily on its own course in a straight line. This is known as a star's *proper motion*. However, some do not behave in this way. Sirius, the brightest star after the Sun, moves in a series of 50-year 'bounces'. This is because it is a binary star with an orbital period of fifty years. Its companion is very faint and was discovered in 1862 as a result of the correct interpretation of the bounces. The centre of gravity of the two stars must move in a straight line; for this to be so it had to be nearer the brighter component than the faint one, and to divide the line joining them in the ratio 2:5.* This shows that the smaller star is only two-fifths as massive as Sirius, so, if we know the combined

mass of the two, we can work out what the mass of each star should be.

What can be found out about a star from the intensity of its light? The brightness of a star is measured by its *magnitude*, a number that decreases with brightness so that a star of the first magnitude is brighter than a second magnitude star, and so on. A difference in one magnitude corresponds to an increase or decrease in brightness by a factor of 2.512, so a first magnitude star is 2.512×2.512 times brighter than a third magnitude star. Originally, magnitudes were estimated by eye, but when astronomical photography was introduced it became much simpler and more accurate to measure star magnitudes from the intensity of their photographic images. The problem was that photographic emulsion was more sensitive than the human eye to blue light, so blue stars appeared to be brighter on a photograph than when viewed directly. As a result it was necessary to specify either *photographic magnitude* (denoted B for blue) or *visual magnitude* (denoted V). Nowadays magnitudes are measured with great precision, typically to a hundredth of a magnitude, using photoelectric devices, and these are fitted with special filters to reproduce photographic or visual magnitudes (or any colour-band you care to specify). The reason that one star appears bluer than another is that it has a higher temperature, so the colour of a star (this can be expressed numerically by subtracting V from B) gives us a direct indication of its surface temperature—another fundamental property of a star.

On its own, the *apparent magnitude* of a star (how bright it appears in the sky) tells us nothing about the star itself, because it could be a faint, nearby star, or a bright, distant one and still look the same. What is more useful is the *absolute magnitude*: how bright it would be if it was at a standard distance, which is chosen to be 10 parsecs (see *The Greenwich Guide to Stargazing*), equivalent to 32.6 light years (l.y.). The absolute magnitude can easily be calculated from the apparent magnitude and the distance of the star,

The path that Sirius (open circles) appears to follow across the heavens is a combination of its uniform motion across the sky and its orbital motion around a much fainter companion, Sirius B (dots). Their common centre of gravity moves in a straight line at a steady speed.

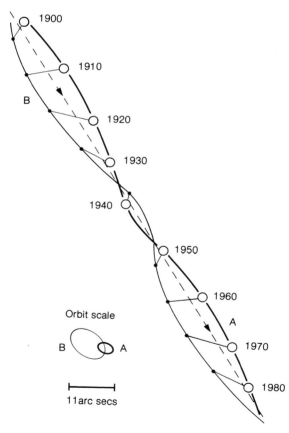

Orbit scale

11arc secs

*This principle can be illustrated by whirling a hammer into the air (make sure there are no people or greenhouses around!). The head goes in small circles while the lighter handle describes bigger circles and the centre of gravity (quite near the head) follows a smooth arc.

and even considering only the relatively few nearby stars whose distances are accurately known, it is clear that there is a huge range of absolute magnitudes. For example, Sirius (absolute visual magnitude 1.4) is 23 times brighter than the Sun (4.8) and 440,000 times as bright as the nearest star, Proxima Centauri (15.5).

There are two reasons why one star should give out more light than another. The first is that it may have a hotter surface—a white-hot star is more luminous than a cooler, red-hot one. Secondly, one star may have a larger surface area than the other. If the Sun were replaced by a star of the same temperature, but twice the surface area, we would receive twice as much light. This gives us a means of estimating the size of a star. If we know its temperature and its absolute magnitude, we can deduce its diameter. Information on a star's diameter can also be obtained by noting how long it takes to fade out as it disappears behind another object (the Moon, or its companion star if it is a binary), but this method can be applied to only a very few stars.

The third way of observing stars, by splitting up their light into its component colours with the aid of a *spectroscope*, has been immensely useful to astronomers and provides the observational basis for the science of astrophysics. The most familiar example of a star's spectrum is a rainbow, in which the light from the Sun is split into its different colours by raindrops. You can also see miniature rainbows cast on the wall when the Sun shines on a piece of cut glass. Pretty though they are, these spectra are of little use because they are too diffuse; to sharpen the image up you need a proper spectrograph in which the light is passed through a slit and focused with lenses. When this instrument was first used by the German optician Joseph Fraunhofer in 1814, he was surprised to find that the Sun's spectrum was not continuous, but was broken by innumerable gaps in the form of dark lines. (They appear as lines because they are images of the slit through which the sunlight is passed.) The positions of these lines in the spectrum correspond very closely with those in spectra obtained in the laboratory by heating certain chemicals to

Light can be broken up into its component colours with a spectroscope. The light source is narrowed into a slit and then passed through a convex lens from which it emerges as a parallel beam. Next comes the prism which disperses the light into its different colours. Finally, a second lens focuses the light on to a screen, or a photographic plate if a permanent record is required.

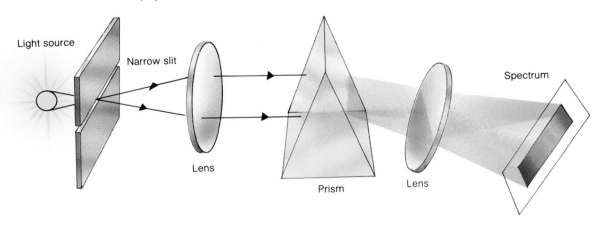

Light source

Narrow slit

Lens

Prism

Lens

Spectrum

incandescence and passing the light through a spectroscope. For example, there are two lines in the yellow part of the Sun's spectrum that are exactly the same colour as those in the light from a sodium street lamp. This is evidence that there is sodium on the Sun.

As a result of matching the lines in the Sun's spectrum to those produced by known chemical elements, a huge range of chemical elements was identified. These included not only the lighter ones like hydrogen and lithium, but also many

heavier ones such as calcium and iron. There remained some lines that could not be identified with any element then known, so they were ascribed to helium (from *helios*, the Greek word for the Sun). Nearly thirty years later, in 1895, helium was found on Earth. The same technique can be applied to stars, by mounting a spectrograph on the end of a large telescope. Like the Sun, most stars are found to contain a large variety of chemical elements.

Knowing what stars are made of is of more value if we also know the relative proportions of the various constituents. This is a much more difficult problem because the intensity of the spectral lines is not a direct measure of the quantity of the element, but depends on the

Five spectra of Vega (Alpha Lyrae) at different exposures. The light from this star, which is three times as massive as the Sun, is separated out into its component colours by means of a spectroscope.

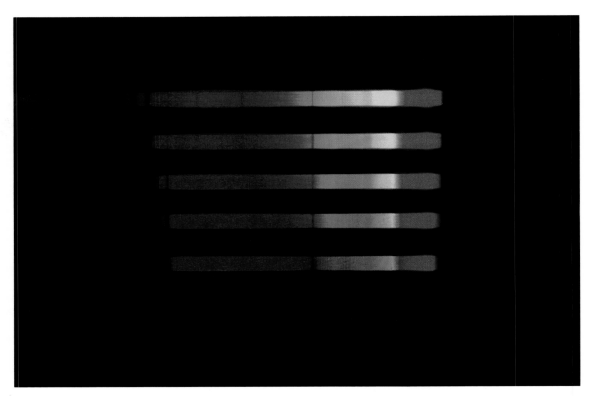

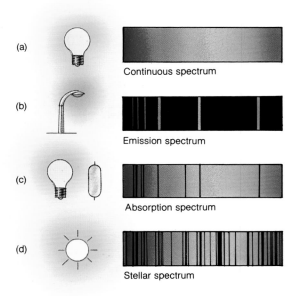

(a) Continuous spectrum

(b) Emission spectrum

(c) Absorption spectrum

(d) Stellar spectrum

Formation of absorption lines:
(a) *A very hot source, for example a light bulb, gives out light at all frequencies;*
(b) *A source like a mercury street light gives out light at only a few, discrete, frequencies;*
(c) *If a hot source shines through a cooler gas, such as a light bulb shining through mercury vapour, the gas will absorb the light of the appropriate frequency, giving dark lines on a continuous spectrum;*
(d) *Similarly, the gases in the atmosphere of a star produce dark lines in the spectrum of light from the hotter, deeper regions.*

temperature and pressure. Also, the lines tell us about the outermost layers of the star where they are formed, so the results may not be representative of the whole star. Nevertheless, a great deal has been learned about how abundant various elements are in stars from a study of their spectra. This information has proved very useful in helping to unravel the life history of stars.

In the early days of astrophysics an attempt was made to classify stars into different types according to their spectra, designated A, B, C etc. It was subsequently decided to drop some of the letters and rearrange the remaining ones in order of decreasing temperature to give the rather peculiar sequence which is now used: O, B, A, F, G, K, M, R, N, S (of which there are very few of types R, N and S). Do not be alarmed if you hear an astronomer muttering 'Oh! Be a fine girl, kiss me right now, sweetheart'; he is probably just reminding himself of the sequence. For greater precision each type is further divided from 0 to 9 in order of decreasing temperature. The Sun, for example, is a G2 star.

Careful measurement of the spectral lines shows that they do not exactly coincide with those obtained from experiments on Earth, but are slightly displaced towards either the red or the blue end of the spectrum. This is due to the doppler effect (see Chapter 1) and the size of the displacement tells us the radial velocity of the star—how fast it is approaching or receding from us. This does not give us any information about the star itself unless, as is not uncommon, the radial velocity shows periodic variations. There are two reasons why this may happen. The star may be expanding and contracting (remember that it is only the near side that we see), or it may be one of a pair orbiting round one another and approaching us in one part of its orbit and receding in another. Stars that are known to be double only because of their changing radial velocities are called *spectroscopic binaries*. When one of the stars is approaching us, its companion will be receding and it may then be possible to see the spectral lines split into pairs, one from each component. Such double-lined spectroscopic binaries are particularly valuable because, from measurements of their radial velocities, we can deduce enough about their orbits to give their relative masses without needing to know their distance from Earth. If, in addition, we find that one star eclipses the other as it goes past (this will show up as a brief change in magnitude at the

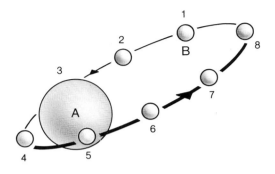

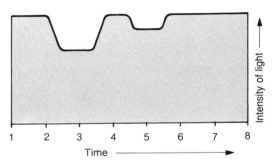

Eclipsing binaries. If a binary star is seen nearly edge-on to its orbit, one component may eclipse the other. This will show up as two troughs in the light curve, one when star B passes behind star A, and another (of different depth) when B passes in front of A.

critical part of the orbit), we will know that the orbit is edge-on to us and can then deduce the mass of each star.

This is not a complete list of everything that can be deduced from the measurement of starlight and no mention has yet been made of observations that can be made outside the range of· visible light (see *The Greenwich Guide to Astronomy in Action*). But it does give a good idea of how optical astronomers set about measuring such fundamental properties of a star as its absolute magnitude, temperature, size, chemical composition and spectral type. Having acquired these data, can we make any sense of them?

We have seen that the absolute magnitude shows a huge range and so does the surface temperature, from as cool as 3000K to tens of thousands of K, with the Sun towards the cool end at 6000K. If you were to plot temperature against absolute magnitude for a large number of stars you might expect to find the points scattered at random all over the graph paper. But when Ejnar Hertzsprung in Europe and Henry Norris Russell in the USA did this independently in 1911 and 1913 respectively, their results were remarkable. With only a few exceptions, the points fell in a narrow band joining 'hot and bright' to 'cold and faint'. This band is known as the *main sequence* and the graph on which it appears is called the *Hertzsprung-Russell diagram*, frequently abbreviated to H-R diagram.

Although over 90 per cent of stars fall on the main sequence, the exceptions are also of great interest. They occur in two distinct groups: one contains white-hot, but relatively faint stars; the other group contains cool, but very luminous stars. For a cool (only red-hot) star to give out so much light, it must have a very large surface area and hence be a very big star: these are the *red giants*. Conversely, the hot but faint stars must be very small and they are called *white dwarfs*.

The horizontal scale of the H-R diagram is temperature, decreasing from left to right. It may be calibrated directly in degrees, or in terms of *B-V*, or according to spectral type: all these are measures of temperature. The vertical scale is absolute magnitude, with faint (large numbers) at the bottom and bright at the top. As we have seen, the size of a star is related to its brightness and temperature, so the stars increase in diameter from bottom left to top right of the diagram.

Soon after stars are formed, they settle down on to the main sequence at a point which is decided by the star's mass, the more massive at the top and the less massive lower down. So long

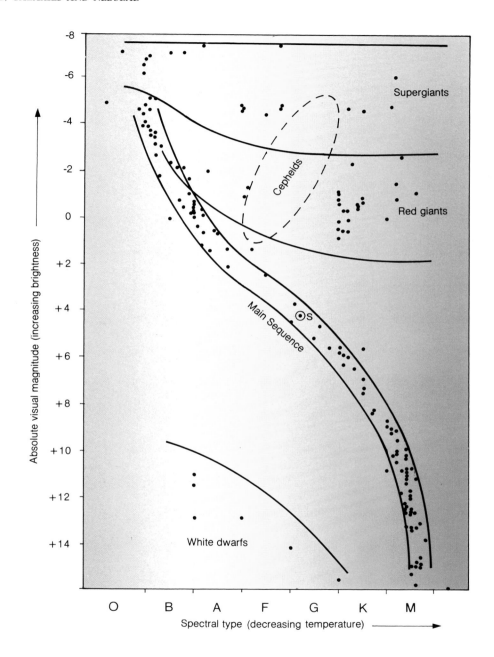

The H-R diagram for the nearest and brightest stars.

as a star continues to produce energy by hydrogen-fusion in its central core (as the Sun does at present), it will remain on the main sequence. But, as we saw in Chapter 2, when the hydrogen in the central core is exhausted and shell-burning commences, the star will become bigger and cooler. This means that it will move from the main sequence towards the right of the diagram. The more massive stars are more prodigal with their fuel supply than the less massive, so stars towards the top of the main sequence start to move right sooner than those lower down; those with less than the mass of the Sun remain on the main sequence for a very long time indeed.

The shell-burning part of a star's life is completed quite rapidly, so very few stars are found at that stage of their evolution and that part of the H-R diagram contains very few points. However, once the core gets hot enough for helium burning, there is enough fuel to keep a star going for a long time. Helium burning is accompanied by a great expansion in the diameter of a star, though its temperature does not change much. This means a vertically upward movement on the H-R diagram as a star settles down to be a mature red giant or, if it is a massive star to start with, as a red supergiant in the top right-hand corner of the diagram. Eventually, old age sets in as the helium runs low. The star starts to collapse and heat up, moving left in the H-R diagram as it does so. The less massive stars sink smoothly and quite rapidly into senility, moving down towards the bottom left of the diagram as hot but very small white dwarfs before finally fading out of sight. But the more massive stars have problems. Before they are able to become white dwarfs they need to lose mass and this they do spectacularly by exploding off their outer layers.

As a star moves around the H-R diagram in the course of its evolution, it might well find itself in a region called the *Cepheid zone*. The qualifications for entry into this zone are only that the star has the appropriate brightness and temperature; age or composition do not matter. The interesting point is that any star within the zone, no matter from which direction it entered, will become a variable star. We saw earlier how, under certain conditions, a star might have difficulty in balancing the rate at which it produces energy with the rate at which it radiates it into space and would pulsate between being a large star with a cool surface and a smaller one with a hotter surface: this is what happens within the Cepheid zone. The name of the zone comes from the star δ Cepheus which behaves in this way, and similar stars are known as *Cepheid variables*. δ Cepheus is easily seen with the naked eye and varies in brightness by about a magnitude over a period of 5.4 days.

Each Cepheid variable has its own characteristic period which may be anything from a little over a day up to fifty days, but interestingly the period is directly related to the absolute magnitude, so if you know one of these quantities you can deduce the other. This makes Cepheid variables very useful as distance-indicators. They are in the top half of the H-R diagram, so they are intrinsically very bright (from a few hundred to many thousands of times as luminous as the Sun) and can be seen without difficulty at very great distances. Because of their variability, they can easily be identified and their periods measured by comparison with neighbouring stars in a series of photographs. Having obtained the period, this indicates the absolute magnitude which can be compared with the apparent magnitude to give the distance. Thus these stars are an invaluable aid in mapping our Galaxy. There is just one catch: Cepheid variables are not the only stars that fall in the Cepheid zone. Some older, Population II stars which show similar periods of variation are also present. These are *W Virginis* stars (again named after the first one to be discovered) and, though they have a similar

period-luminosity relationship to that of the classical Cepheids, they are systematically over a magnitude fainter. This means that they would indicate the wrong distance if mistaken for a Cepheid variable.

Another type of Population II variable star is the *RR Lyrae* variable near the bottom of the Cepheid zone, of which some 6000 have been found. These have periods of less than a day, but they all have the same absolute magnitude: about 0.6. These are useful distance-indicators for Population II objects, such as globular clusters (see p. 46), and have also been invaluable for mapping our Galaxy. Because they are much fainter than Cepheids, RR Lyrae stars are not so easy to see at great distances.

There are many other types of variable star. Some, such as eclipsing binaries and exploding stars (novae and supernovae), have been mentioned already. More accessible are the quite common *Mira variables* (over 5000 are known), which are particularly suitable for observation by amateurs. Mira itself is also known as O Ceti and is a giant red star over 400 times the diameter of the Sun, towards the top right of the H-R diagram. Its variation is irregular, which is why continuing observations are required, but typically it varies between a comfortable naked-eye magnitude of 2.5 down to magnitude 10 and then back again over a period of about 11 years. Other Mira variables have periods ranging from a month to three years and magnitude ranges of between 1 and 11, though none of them repeats accurately from one cycle to the next. All such stars have a huge, extended surface and, as a loose analogy, it may be useful to visualize them as giant, oscillating soap bubbles, though omitting the final pop!

The constellation Cassiopeia, which is shaped like a W with its base towards the bottom right. The images are 'trailed' slightly because of the rotation of the Earth during the exposure. It is easy to distinguish the hot, blue stars from the cooler red ones.

4 · *Fuzzy Objects in Space*

With the exception of the Sun, all stars, even the most vast, are at such an enormous distance from us that they can be seen only as points of light, even through powerful telescopes. But there are many objects in the night sky that are more rewarding to view, appearing as fuzzy blobs.

They are all known as *nebulae*, from the Latin word for mist. But this is a very superficial description and covers a huge range of widely different structures, in the same way that grouping together all the books in a library that have red covers tells us what they look like from the outside, but gives us no clue about what they contain. The main division is between objects that are part of our own Galaxy and so relatively nearby, and objects outside our Galaxy—the extragalactic nebulae. As we saw in the first chapter, this means that they are themselves galaxies, consisting of thousands of millions of individual stars, but so remote that they appear only as blurs in all but the most powerful telescopes. Because of their great importance in astronomy, they are given their own chapter. Of those within our Galaxy, the principal division is between the true nebulae, i.e. gaseous objects, and bodies which appear fuzzy at low magnification but which resolve themselves into star clusters on closer examination.

These fuzzy objects were first catalogued by the French astronomer, Charles Messier, over 200 years ago, so that he would not confuse them with the comets he was seeking. The Messier objects he listed are designated by M (for Messier) followed by their catalogue number, but his index of just over 100 entries was superseded by a list of over 5000 published in 1833 by Sir John Herschel (son of the discoverer of the planet Uranus), and this in turn was superseded in 1888 by the *New General Catalogue of Nebulae and Star Clusters* produced by Johann Dreyer, a Danish astronomer who worked in Ireland. Dreyer later published two *Index Catalogues* bringing the total number of listed objects to over 13,000. Nebulae are referred to by the initials of the catalogue (M, H, NGC or IC) followed by the entry number, which can lead to some ambiguity when an object appears in more than one catalogue. Moreover, many of the brighter objects also have more popular names. For example, M8 and NGC 6523 are both alternative names for the Lagoon Nebula, M1 is better known as the Crab Nebula and M42 as the Great Orion Nebula.

Star clusters

Although they are not true nebulae, star clusters are of considerable interest. This is partly because they are good to look at, but also because of what they can tell us about astronomy. Star clusters come in two types: open (or galactic) and globular. *Open clusters* are groups of up to a thousand stars set close together in space, but in no particular arrangement. Several open clusters can be seen with the naked eye, such as the Hyades near the star Aldebaran, the eye of the bull in the constellation of Taurus. Better still are the Pleiades, also in Taurus, but slightly further

The gas clouds of NGC 2359 are illuminated by hot, bright stars within them.

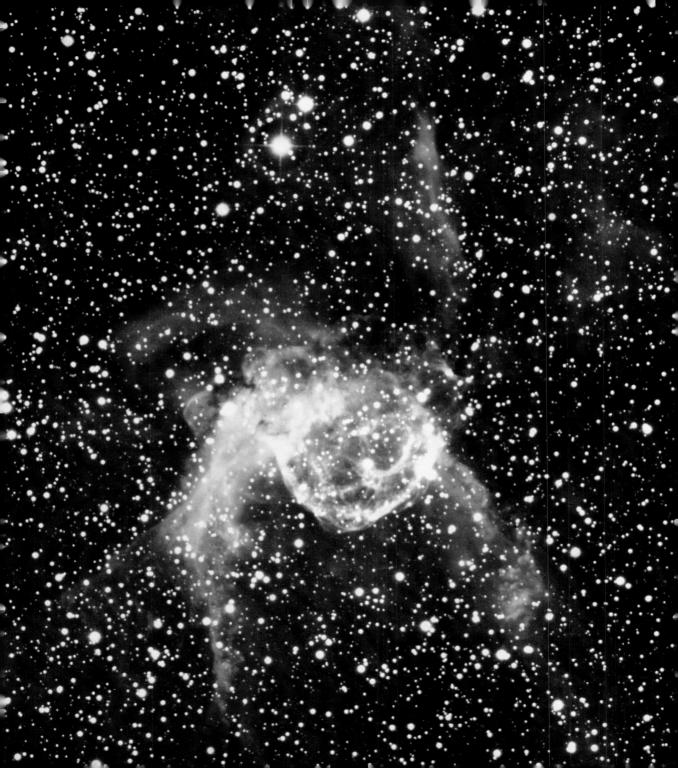

to the north-west. For observers in the southern hemisphere there is a fine but more compact example known as the Jewel Box around κ Crucis in the Southern Cross. Many other examples can be seen with binoculars. At first glance the Pleiades (named after the seven daughters of Atlas who turned into doves as they fled from Orion) appear only as a patch of light to

Some of the stars in the Pleiades open cluster. These are very young stars and the nebulous material around them is the remains of the gas cloud from which they are condensing.

The open cluster NGC 6520 seen against the background of the Milky Way, whose yellowish stars cover most of the photograph. The cluster stars are mostly hot and blue, but although it is young, some of its stars have already evolved into red giants. In the lower half of the photograph the view is obscured by a dust cloud, which appears as a dark patch.

the naked eye. Also at second glance to me, but most people can see five separate stars and those with particularly good eyesight can pick out all seven. These are just the brightest stars of a group which includes over a hundred stars.

The beauty of star clusters for the astronomer is that, because all the stars in one cluster are close together in space, they almost certainly all formed out of the same gas cloud at about the same time, so all the stars are the same age and had the same initial composition. The only

45

difference between them is that some are bigger than others. Imagine what an advantage it will be to a visiting Martian, trying to understand the life-cycle of human beings, when he discovers that there are schools where all the children are conveniently divided into classes according to their age and where all have come from similar backgrounds. The analogy does not stop there because, like schoolchildren, the stars in open clusters are all young compared with the averge population. The gravity of an open cluster is not sufficient to hold it together for more than 100 million years, so the examples we see cannot be older than this. The H-R diagrams for individual clusters confirm this relative youth, showing that even the most massive members have not evolved far from the main sequence. Also, long-exposure photographs of open clusters frequently show remnants of the original gas clouds from which the stars formed; again, the Pleiades are a good example.

It is particularly easy to construct H-R diagrams for clusters because the most difficult measurement of all, the distance, is the same (near enough) for all the stars in the cluster. This means that the difference between the absolute and apparent magnitude is the same for each star, so you can get the right picture by plotting colour against *apparent* magnitude. Although the numbers on the vertical scale will be wrong, they will all be out by the same amount. Indeed, you can go a stage further by sliding your un-calibrated H-R diagram up or down until the lower end of the main sequence coincides with that on a calibrated diagram. This will give you the correlation between apparent and absolute magnitude, from which you can calculate the distance of the cluster.

Globular clusters, of which about 200 examples are known in our Galaxy, are not at all the same. Easy naked-eye examples are M13 in Hercules for northern observers, or ω Centauri for those in the southern hemisphere (see *The Greenwich Guide to Stargazing*), but their true beauty can be

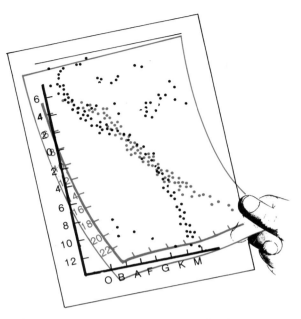

The H-R diagram of a cluster (red) can be slid up and down until the lower part of the main sequence coincides with that of a calibrated H-R diagram (black).

appreciated only with binoculars or a telescope. They consist of hundreds of thousands of individual stars forming a sphere, with the density of stars increasing towards the centre. Globular clusters have a rather curious distri-bution in space. Whereas most of the stars in our Galaxy (including the open clusters) form a flattened disc whose centre is the middle of the Galaxy, the globular clusters are distributed, somewhat like the stars they contain, in a sphere or halo, with the density greater towards the middle. Also many of them are well away from the galactic disc. The centre of this globular cluster of globular clusters is the centre of the Galaxy.

The globular cluster 47 Tucanae.

H-R diagrams for globular clusters can be produced in just the same way as for open clusters, by assuming all the stars of one cluster are at the same distance and using apparent magnitude instead of absolute magnitude. But the resulting diagrams are quite different, with all the brighter stars lying well away from the main sequence. Globular clusters are very old indeed. In fact, they are Population II stars, the first generation of stars to form after the Big Bang. Because stars are closely packed towards the centre of globular clusters, they are strongly bound together by gravity and have survived into old age without dispersing.

Globular clusters have been of great value in investigating the size and mass of the Galaxy because some of them are the most distant objects that can be seen in the Galaxy. Their positions can be completely specified by their RA, declination, and distance, the last being obtained from the magnitudes of RR Lyrae stars (see p. 40), which are usually to be found in globular clusters. In addition to their positions, we can also measure one component of their velocity—the line-of-sight, or radial velocity—by using a spectrograph to measure the doppler effect. To avoid falling into the centre of the Galaxy, globular clusters must be in orbit around it, just as the planets orbit the Sun. The speed of a planet in its orbit depends on its distance from the Sun and the mass of the Sun. If we know the speed and distance, we can deduce the mass of the Sun. Similarly, the speed of globular clusters and their distance from the centre of the Galaxy can be used to estimate the mass of the Galaxy.

It is not quite that straightforward, because we know only one component of the velocity and

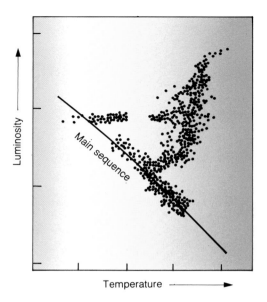

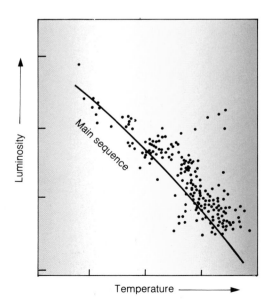

The H-R diagram of the globular cluster M3 (above) shows that most of its stars have evolved away from the main sequence, indicating great age. In contrast, most of the stars of the open cluster NGC 2264 (below) are still close to the main sequence, so it must be very young.

have no reason to suppose that the orbits are circular, but with statistical arguments it is possible to obtain a good estimate of the mass of the Galaxy. What is embarrassing is that the figure obtained in this way is far greater than that which would be expected from summing the masses of all the visible stars. The 'missing mass' must come from objects that give out no light, and there is considerable speculation about what these can be. Some scientists think that there may be vast numbers of black holes (see p. 87) in the Galaxy; others propose even more extreme solutions involving dramatic new ideas such as cosmic strings (very thin, incredibly heavy strands), or even vast quantities of a new type of particle that has yet to be discovered, called a *wimp* (a 'weakly interacting massive particle'). Perhaps one of these will prove to be the correct answer or maybe they are all wrong. The one certainty about the missing mass is that it presents a problem which will keep theoretical astronomers happily occupied for many years to come.

Planetary nebulae

Planetary nebulae are objects which, when viewed through a telescope, appear as small circular discs of light which look somewhat similar to planets—hence the name which was given to them by Sir William Herschel. However, it is obvious that they are not planets, firstly because they do not move relative to the stars and secondly because stars can be seen within the disc. If these stars are between us and the disc, the nebula must be further away than the stars and much, much bigger, so it cannot be a planet. If the stars are beyond the disc, it must be transparent, so again it cannot be a planet.

In fact, planetary nebulae are shells of gas that have come from a giant star which is nearing the end of its existence and is evolving into a white dwarf. Small stars like the Sun can, when their time comes, evolve peacefully from giants to white dwarf stars without any fuss. But Subrahmanian Chandrasekhar, the Nobel Prize-winning astronomer, has shown theoretically that the mass of a white dwarf cannot exceed 1.4 solar masses, because the material of which it is composed would collapse under the pressure of any greater mass. So what becomes of a more massive star when it runs out of fuel? The really big ones turn into neutron stars or black holes, which will be discussed in Chapter 6. Those that do not greatly exceed 1.4 solar masses can still become white dwarfs, but have to have a crash slimming course first to get their mass below the limit. Crash is the right word; the excess mass is disposed of by flinging off the star's outer layers at a speed greater than the escape velocity (otherwise they would fall back again). These form an ever expanding gas shell around the star. The new, slimline star at the centre of the shell can now become a white dwarf, with the characteristic property of being very small and very hot; so hot that much of its radiation is in the form of ultra-violet light. As the ultra-violet light passes through the shell of gas, some of it is absorbed and then re-emitted as visible light and that is what we see as a planetary nebula.

The light from planetary nebulae is interesting. When it was analysed by passing it through a spectroscope, it was found to include green lines that could not be identified with any known chemical element. The early spectroscopists thought they had discovered a new element which they called 'nebulium', but it is now known that the light comes from oxygen. At ordinary pressures like those found on Earth, the green line in the oxygen spectrum is undetectable, but at very low pressures like those found in nebulous gas, the green line continues to shine while the other stronger ones fade out. It would still not be detectable in a small sample of gas, but can be seen in something the size of a planetary nebula. Indeed, through a large telescope it is often possible to identify planetary nebulae purely because of their green colour—there are

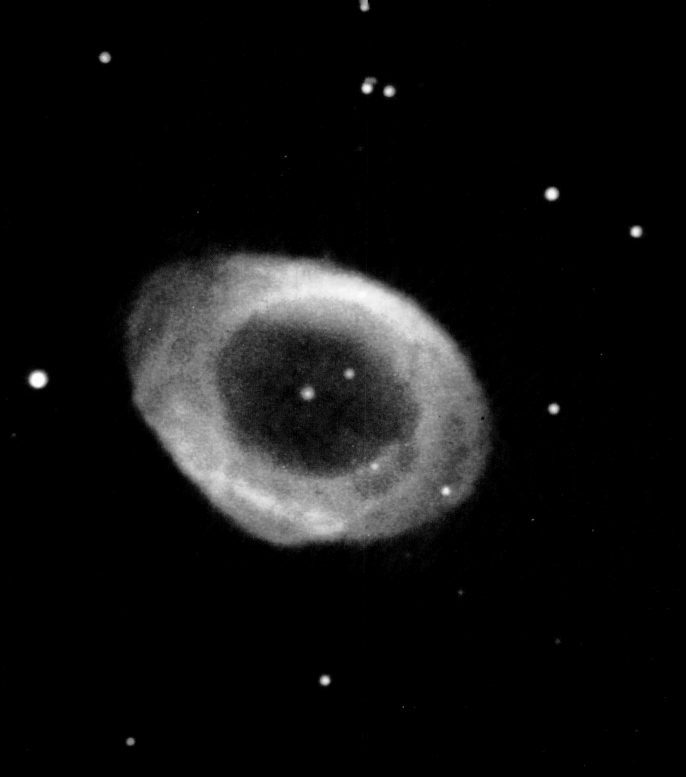

LEFT *The Ring Nebula in Lyra, NGC 6720. This planetary nebula is a spherical shell of glowing gas, half a light year across, surrounding a hot central star.*

ABOVE *The Helix Nebula, NGC 7293. This planetary nebula's expanding shell of gas is much less uniform than that of the Ring Nebula.*

51

The Dumbbell Nebula in the constellation Vulpecula. Although the expanding gas shell is not a complete sphere, this is clearly a planetary nebula.

no green stars—although many planetary nebulae show other colours as well as green, so this method is not infallible.

Not all planetary nebulae are perfect discs. Many of them show a more complicated structure, either because the explosion that formed the shell was not symmetrical, or because the shell was thrown off, not all at once, but a layer at a time. Also, although we are confident that there must be a central star, or the shell would not glow, it is not always possible to see one. This is not too surprising, as white dwarf stars are of very low magnitude, despite their high temperature. Some central stars are visible, however, and these account for about 20 per cent of all known white dwarfs.

Since all stars between 1.4 and about 4 times the mass of the Sun are thought to pass through the planetary nebulae stage, we might expect them to be quite common objects and about a thousand examples are known. But they do not last long in astronomical terms. The gas shell is continuously expanding at a few tens of kilometres per second and it does not get replenished, so after about 100,000 years it will be too dispersed to be seen.

Gaseous nebulae

These are among the most beautiful of all objects in space, but they have a very low surface brightness and are revealed in their true glory only in very long-exposure photographs taken with large telescopes. Nevertheless, some of them are well worth looking at through binoculars or a telescope providing you do not expect to see too

The Orion Nebula can easily be seen with the naked eye in the middle of Orion's sword, but an image of this quality requires a powerful telescope and the extraordinary photographic-processing skills of David Malin.

much. The most famous example is the Great Nebula in Orion (M42), which appears to the naked eye as a small fuzzy patch in Orion's sword.

As the name suggests, *gaseous nebulae* are simply tenuous clouds of gas. There is a considerable amount of gas about in the Galaxy, but it is not uniformly spread, being concentrated towards the centre of the Galaxy and largely found in its spiral arms (see Chapter 5). It consists mostly of hydrogen and does not normally give out any visible light, because it is too cold. Much of our knowledge of it comes from astronomical measurements using radio waves, as these are emitted at temperatures far lower than those required for the emission of visible light. The

gaseous nebulae that show up in telescopes do so either because they are reflecting starlight or because, like planetary nebulae, they are absorbing energy emitted by stars and re-emitting it as visible light. This can only happen if there are bright stars nearby, which is often the case as it is from just such clouds of gas that stars are formed. These two types of nebulae are known as *reflection nebulae* and *emission nebulae*, though it is not unusual to have both processes in the same nebula.

A third type of gaseous nebula is the *absorption nebula*, which shows up only because it blots out the background stars leaving a dark patch in the sky. By far the best example of an absorption nebula is the Coal Sack, in the southern constellation Crux. This shows up particularly well because it is in line with a very rich part of the Milky Way where the absence of background stars is very noticeable. Other evidence of absorbing material comes from detailed study of the spectra of stars. Some of these show absorption lines which cannot originate in the star because their doppler-shift is quite different from that of the star. They result from the absorption of some of the starlight by interstellar gas that lies between us and the star.

LEFT *The Cone Nebula in Monoceros. The characteristic dark cone is produced as a result of the sheltering effect of dust clouds, which shield this region from the radiation produced by the hot central star.*

OVERLEAF LEFT *The Horsehead Nebula in the constellation of Orion. The dark feature that gives it its name is a dust cloud that obscures the view of the bright nebula behind it. The gaseous nebula NGC 2023 at bottom left shines by reflecting starlight.*

OVERLEAF RIGHT *The Trifid Nebula in Sagittarius. Its hydrogen gas clouds are illuminated by recently-formed hot stars within it, but some of the glow is obscured by irregular bands of dust.*

The most spectacular gaseous nebulae are those where there is a concentration of gas near a hot star. This can occur either at the start of a star's life, as in the nebulosity around the young Pleiades, or towards its end, as in the case of planetary nebulae around aged stars. The Orion Nebula is another example of a young nebula, where stars are still forming and where much of the energy that excites the glow from the nebula comes from a proto-star within it. The red light of the nebula, seen only in long-exposure photo-

The Eta Carinae Nebula, a fine gaseous example in the southern Milky Way. Ultra-violet radiation from the massive star within it, Eta Carina, provides the energy which causes the hydrogen clouds to glow.

graphs, is characteristic of hydrogen gas. Other examples of nebulosity around aged stars are the remnants of supernovae—stars that have blown themselves to pieces—such as the Crab Nebula and the Tarantula Nebula, and these will be looked at in more detail in a later chapter.

5 · Beyond the Galaxy

Extragalactic nebulae is the name given to those nebulae that are outside our own Galaxy. In fact, they are themselves galaxies not unlike our own; it is only their huge distance that makes them appear to be of less significance than objects closer to home. As we saw in the first chapter, groups of galaxies are the building blocks of the Universe, and much of our knowledge of the origin and evolution of the Universe has come from studying their distribution and movements. From measurements of the doppler effect we know that they are moving with very great speeds relative to one another—some are receding from us at speeds approaching the velocity of light and there is no reason to doubt that their sideways movements are equally large, though their distances from us are so great that we can never hope to detect any changes in their positions in the sky. This in itself is useful to positional astronomers, as they can use the background of the galaxies to provide a fixed frame of reference against which to measure the movements of the stars. Although this has been recognized for some time, it is only recently that it has been possible to measure galaxies' very faint and slightly fuzzy images accurately enough to take advantage of their stability.

Just as the Sun, as an average star, was a convenient starting point for looking at stars, so our own Galaxy, as a fairly typical example, might be a good starting point for galaxies. We are certainly close enough to see it in some detail, but in some ways we are too close. Since we are actually within the Galaxy, it is difficult to see the wood for the trees. To get an overall view of what a typical galaxy looks like, it is better to look first at one of the other galaxies in our local group, the Andromeda Nebula (M31), which is almost a twin of our own, though a little larger.

The Andromeda Nebula can be seen with the naked eye as a small, faint smudge of light. Through binoculars the galaxy is clearly elliptical, but it requires a long-exposure photograph to show it up properly. The reason it appears to be elliptical is that we are looking at it at an angle. If we could view it square on, it would be approximately circular. Viewed edge-on it would be cigar shaped. It contains about a hundred thousand million stars, but they are not uniformly distributed. The greatest concentration is towards the middle, which appears much brighter than the rest (this is the part that can be seen with binoculars). Around this central hub are clouds of stars arranged in loose strings that appear to spiral out from the centre. These are the *spiral arms*.

Now we know what to expect, we are in a better position to interpret our own Galaxy. Whichever way we look in the sky we can see stars, but in some directions they are more numerous than in others. In particular, there is a band round the sky known as the Milky Way which is particularly rich in stars. The milkiness is the combined light of huge numbers of distant stars. In contrast, the direction at right-angles to the Milky Way has very few stars. By analogy with the Andromeda Nebula, it is clear that when we are looking at the Milky Way we are viewing along the plane of the Galaxy, and at right-angles to it we are looking out of the Galaxy. The Milky

The Great Andromeda Nebula is a spiral galaxy very like our own. The straight line in the photograph was caused by a passing satellite.

Way itself is not uniformly bright. The most spectacular view of it is from the southern hemisphere, looking towards the constellation of Sagittarius. In comparison, the northern Milky Way is insipid. Again, the reason is obvious: Sagittarius is in the direction of the centre of our Galaxy, whereas the northern view is towards the rim. Our position is in one of the spiral arms towards the outside edge of the Galaxy.

Although the Milky Way is most intense around Sagittarius, the band of light in that direction is broken up by darker patches of dust and gas which are obscuring the view. This is unfortunate, because it means that we cannot see the centre of the Galaxy, or anything else that is close to the plane of the Galaxy. Instead we have had to piece together our knowledge of the structure of the Galaxy from radio and infra-red astronomy. This is because radio waves and infra-red light are better able to penetrate the gas and dust that absorbs visible light, in the same way that your radio continues to work even when it is cloudy.

This wide-angle photograph shows how the density of the Milky Way increases towards Sagittarius. But the galactic centre itself is hidden from view by dust in the spiral arms.

Radio emission is picked up from many parts of the Galaxy, including the galactic centre, and with its aid astronomers have been able to trace the spiral arms of our Galaxy. We are on the outer edge of one of them and there are another two between us and the centre of the Galaxy. There is also another one further out at a distance of about 40,000 l.y. from the galactic centre and 8000 l.y. from the Sun. All the spiral arms are approximately in the same plane, forming the *galactic disc*. But this represents only a few per cent of the total mass, most of which is concentrated in the bulge around the galactic centre. In addition, as we saw in the last chapter, there is a halo of Population II stars, mostly in globular clusters, in a sphere which extends out even beyond the spiral arms, though it is much more densely populated towards the centre. These three parts, the central bulge, the galactic disc and the halo, make up the Galaxy.

Radial velocity measurements show that the Galaxy is rotating. This is to be expected, not only because it looks as if it is rotating from its shape, but mainly because, if it were not rotating, everything would have collapsed towards the centre long ago. Just as the planets orbit the Sun under the influence of its gravity, so all the components of the Galaxy are in orbit around its centre. By analogy with the planets, we would expect the parts nearer the centre to complete an orbit in less time than those further out. It is tempting to try to explain the spiral arms in this way. If the Galaxy had started like a rimless cartwheel, with all the spokes pointing straight out from the centre, then the longer orbits of the stars on the outer parts of the spokes would cause them to lag behind those nearer the centre, curving the spokes into spirals. This gives the right direction for the rotation of the Galaxy with the convex side of the spiral arms leading.

But the explanation is not that simple. We have reason to believe that the Galaxy is at least 10 thousand million years old, because some of its stars are that age. In that time the outer parts, such as those in the vicinity of the Sun, would have completed about fifty orbits and the inner parts at least a hundred more. This would mean that the spirals should each have about a hundred turns and the Galaxy should look more like a gramophone record than a bent helicopter blade. Also, observations of other galaxies show that the rotation is not quite like that of the planets. In a planetary system the outer parts take longer to complete an orbit than the inner ones. Another form of rotation is like a rigid wheel where all the parts take the same time to complete a revolution. Galactic rotation is a compromise between the two, becoming more wheel-like and less planet-like as you go towards the centre.

So what has caused the spirals? Various ideas have been put forward, but the most likely is that they are *density waves*. These regions of high density are formed when the orbiting stars encounter a sort of traffic jam, which causes temporary bunching in the same way that roadworks will cause a 'density wave' of traffic to build up behind them. The movement of the wave is not the same as the movement of the stars, which will pass through a density wave just as cars pass through a traffic jam. The beauty of density waves is that, besides explaining the observed spiral structure of other galaxies as well as our own, they also provide a suitable triggering mechanism for the formation of stars out of gas clouds. It is within the spiral arms that stars are seen to be forming.

The galactic bulge, like the halo, contains very old, Population II stars. Indeed, the halo and the bulge together are somewhat like a giant globular cluster. It is only within the spiral arms that the second-generation, Population I stars are to be found. Towards the centre of the Galaxy

The structure of our Galaxy, edge-on and square-on. The central bulge is the nucleus, with globular clusters forming a three-dimensional halo round it. The spiral arms are confined to a thin disc.

The spiral galaxy NGC 2997.

the density of stars gets very great, and evidence is accumulating that right at the centre there is one of the most curious things in the Universe—a massive black hole. Any black hole is remarkable enough and we shall learn more about them in the next chapter, but the one believed to be at the centre of the Galaxy is outstanding even amongst this peculiar breed. It is thought to be as massive as three million Suns!

Our own Galaxy is typical of thousands that are to be found in the Universe. Some, like the Great Nebula in Andromeda, are larger, while others are smaller. The number and complexity of the spiral arms also varies, but these spiral galaxies all have the same essential features of bulge, halo and spiral arms. It is possible that they all have central black holes, but it would be safer to await confirmation of our own black hole before making this assumption. Much of the variety in their appearance is the result of the

angle from which we view them. Seen square on they look like Catherine wheels, viewed from the edge (as we view our own Galaxy), the obscuring gas and dust in the galactic plane forms a dark band with the bulge showing above and below, just like a miniature picture of the Milky Way in the direction of Sagittarius. From an artistic point of view, the most satisfying spiral galaxies are those seen from an angle, like Andromeda.

One feature that many spiral galaxies show, but our own does not, is a bar. The characteristic feature of a *barred spiral* is a pair of straight arms sticking out from the bulge, with a pair of spiral arms tagged on the end of them so that the overall shape is somewhat like an 'S'. Like the

The Sombrero Galaxy in Virgo, a spiral galaxy seen nearly edge-on.

OVERLEAF *In contrast to the Sombrero Galaxy, Messier 83 is nearly square-on. This photograph shows clearly how the spiral arms are distributed around the central nucleus.*

spiral arms, this bar can be explained in terms of density waves. Whether a galaxy develops a bar or not, according to the theory, depends on how much mass is in the halo. While this works very well for computer models, it is very difficult to test the theory by observation because of the problem of measuring the mass of a halo. Anyone who has looked at galaxies through a powerful telescope will understand why. These are extremely faint and distant objects and it is remarkable that we are able to learn anything about them at all.

There is a large variety of barred spirals,

The spiral galaxy NGC 300 fills the photograph, though the outer parts of the spiral arms are not immediately obvious.

ranging from those with hardly any bar at all to others where the bar is the dominant feature. There is no reason to believe that barred spirals are essentially different from ordinary spirals; they are all members of the same family. In any case, although it is the arms that catch the eye, they represent only a small part of the mass of the whole galaxy.

Not all galaxies are spirals. There are also *elliptical galaxies*, *lenticular galaxies* and *irregular galaxies*. Elliptical galaxies are, as the name suggests, either round or elliptical. The stars in them are very regularly distributed, giving a uniform glow which fades towards the edges. But they show a huge variation in size and mass. Most are smaller than our Galaxy, but some giant ellipticals are much larger, with sizes ranging from as little as one-thousandth of the mass of our Galaxy to over a hundred times as massive. We see them as two-dimensional objects, but it would be interesting to know what shape they are in three dimensions. At one time it was thought that they were elongated like an American football, but they could equally well be squashed spheres, like a partly deflated beach ball; there is no way of telling just from the appearance. Indeed, they might all be the same shape, with the apparent difference in elongation resulting from the angle from which we view them, just as spiral galaxies can appear circular, elliptical or cigar-shaped depending on the point of view. However, a straightforward count of the ellipticals with different elongations shows that this is not the case. Assuming that they are randomly orientated, we would not expect such a high proportion to appear nearly circular.

If a soggy sphere is spun, it spreads out along its equator and contracts along its spin axis, so, if the galaxies are rotating, the squashed sphere model looks promising. This can be tested, though not without difficulty, by measuring radial velocities on opposite sides of the bulge. It is not possible to do this for individual stars, of course, but an average velocity can be obtained from the combined light of millions of stars. When this was done, it was found that the galaxies were not rotating fast enough to explain the amount of flattening. It seems that elliptical galaxies are neither squashed nor elongated spheres, but would appear as an ellipse (though with different amounts of elongation) from whichever angle they were viewed. In geometrical language, they are ellipsoids and not spheroids.

As well as giving the radial velocities, spectra of elliptical galaxies can tell us about their average temperature and composition. In general, they contain a high proportion of old, cool, red giant stars, but they are also rich in metals. For this to be so, the evolutionary process must have occurred more rapidly than in spiral galaxies like our own. They have gone through all the stages of star formation, metal enrichment and ageing, and the star orbits have been evened out, but they still have some surprises left. Some ellipticals are found to contain gas and others dust in the form of a disc, and many of these are powerful radio transmitters. We are still a long way from understanding these intriguing galaxies.

Lenticular galaxies are so called because they are lens-shaped, with a bulge and a disc. They look somewhat like spiral galaxies without the arms, or, alternatively, like overstretched ellipticals. Astronomers who classify the different types of galaxies have tended to use the designation SO (lenticular galaxies) as a sort of rubbish bin for those that do not fit easily into any other category. A more careful re-classification, using better images, would certainly put some of them among the ellipticals and others among the spirals, but many would still be left over. So though lenticulars have similarities to both spiral and elliptical galaxies, they do form a class of their own.

In overall appearance, they are more like

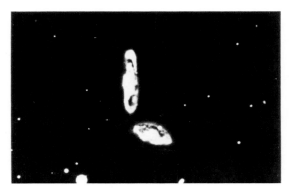

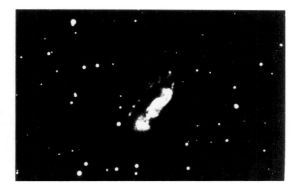

A selection of peculiar galaxies (clockwise, from top left): *NGC 4656/7 in Canes Venatici; NGC 2545 in Cancer; NGC 6621/2 in Draco; and NGC 3986 and 3988 in Ursa Major.*

limbless spiral galaxies, because of the disc. But the stars they contain are more like those found in ellipticals, mostly old and cool. There is no evidence of any recent star formation in lenticular galaxies. The one clue about the origin of lenticulars is that they tend to be more common in dense clusters of galaxies than in more open clusters. This suggests that the proximity of other galaxies has something to do with their formation.

Irregular galaxies have no particular struc-

ture, but are simply chaotic assemblies of stars. They are pretty small objects by galactic standards, no more than a few thousand million solar masses. However, they are the easiest extragalactic nebulae to see, because two of them, the Magellanic Clouds, named after the Portuguese explorer Magellan, are particularly close to us. They are easy naked-eye objects in the southern sky, and look very much like small puffs of cloud. Indeed, on many occasions I have thought that I was in for an early night because the sky was clouding up until I realized that it was only the Magellanic Clouds. To distinguish between them,

The irregular galaxy NGC 6822, which is a member of our local group.

they are known as the Large and Small Magellanic Cloud, abbreviated to LMC and SMC. The LMC is nearer to us (at about 150,000 l.y.) than the SMC (200,000 l.y.), but the difference in size is not just the effect of distance; the LMC really is the larger.

To get some idea of relative sizes and distances, imagine Australia as our Galaxy (with Ayers Rock, appropriately enough, as the mysterious black hole), Borneo as the LMC and Thailand as the SMC. On this scale the Andromeda Nebula would be a sixth of the way to the Moon. These three galaxies and about twenty others are part of our *local group*, but the LMC and the SMC are easily the closest, so close in fact that they have an appreciable interaction with our Galaxy.

The Large Magellanic Cloud, a nearby irregular galaxy which is a prominent feature of the southern sky.

The Tarantula Nebula in the Large Magellanic Cloud, also known as 30 Doradus. This huge gas cloud has the mass of half a million Suns, though thinly spread over a diameter of 700 light years, with nebulous filaments stretching much further. Ultra-violet radiation from a group of very hot luminous stars at its centre makes it glow so brightly that it can just be seen with the naked eye, even though it is a hundred times more distant than the Orion Nebula.

Radio astronomers have detected a bridge of gas between us and the LMC, presumably drawn out of the LMC by the gravitational attraction of our Galaxy. There is also a bridge of gas between the two Magellanic Clouds.

Like most irregular galaxies, the Magellanic Clouds are rich in gas and poor in heavy elements. This suggests that evolution has been rather slow there. Star formation is still going on and appears to take place in localized bursts

NGC 55 in Sculptor is something of a puzzle. Is it an irregular galaxy, or a spiral galaxy seen edge-on?

rather than uniformly throughout the galaxies. One such locality is the Tarantula Nebula in the LMC. This region, with its complex clouds and young stars, is very like a type of irregular galaxy identified in the 1950s by Fritz Zwicky, known as blue compact galaxies. Although very small (a few tens of millions of solar masses), these are of particular interest because they have evolved so little and give us a glimpse of the primordial material from which all galaxies were formed soon after the Big Bang.

In the world of extragalactic nebulae, it is the spiral nebulae that steal most of the limelight. As they are very photogenic and we live in one, this is not surprising. They also appear to be the most numerous kind of galaxy, but this is an illusion. The other types tend to be less luminous, but a careful count shows them to be more common than spirals. We have learned quite a lot about processes in spiral nebulae from studies of our own, but the other types provide a rich, if difficult, field of study of evolutionary processes in conditions very different from our own.

It would be satisfying to end this chapter with a unified theory of galaxies, with some sort of galactic H-R diagram to show how all the different types fit in and how one evolves into another. But unfortunately this cannot be done. Sometime, soon after the Big Bang when galaxies first started to form, some were destined to become elliptical, others spiral, and others lenticular, but we do not know why. We know that they do not evolve from one type into another, because there simply has not been enough time. The lifestyle of a star is mainly determined by its mass, but this is not the case with galaxies; although spirals tend to be larger than ellipticals, both sorts show wide and overlapping ranges of mass. Other factors such as spin, magnetic field, interaction with other galaxies and anything else you care to think of may be involved, but at present we do not know. Not only do galaxies stretch our observing ability to the limit, they also stretch our minds.

6 · Curiouser and Curiouser

As any zoo-keeper will tell you, it is the rare and exotic animals that attract the most attention from visitors, and the more outlandish the better. Who wants to look at a cow when they can see a giraffe? It is the same with astronomy; who wants to look at a common type of star when they can see a rare one?

First, there are radio sources. These are not necessarily peculiar objects: indeed, the first radio source to be identified was the Sun, a very ordinary star indeed! But radio astronomy still has a certain novelty because it was developed only fairly recently, in the late 1940s, although its origins can be traced back to 1931. It is not surprising that it did not start earlier. Even if Galileo had had access to a sensitive, directional radio receiver, he may well not have thought of pointing it at the sky.

Radio astronomy started in 1931 with an attempt to trace the source of radio interference. The scientist concerned—Karl Jansky—found to his surprise that it was coming from the direction of the Milky Way. (In recognition of this pioneering work, the unit of intensity for radio sources is called a jansky.) The systematic study of radio astronomy was delayed by World War II, though the wartime development of radar and electronics proved to be just what was required for this new science. Since then most of the instrumental work has been directed towards producing increasingly sensitive detectors, and to developing means of measuring the direction of radio sources with ever-increasing precision. After an initial burst of cataloguing radio sources in the heavens, attention turned to making intensity-contour maps of interesting objects—the radio equivalent of photographs.

Except that our eyes cannot detect them, the only essential difference between light and radio waves is that the radio waves have a much longer wavelength. They travel at the speed of light and, like light, are a form of electromagnetic radiation. Also like light, radio emission contains a range of different frequencies, corresponding to the spectrum of visible light. It has the great advantage that it can penetrate clouds and dust so a radio astronomer can work on a cloudy night, and can also 'see' through much of the gas and dust that obstructs the view of the optical astronomer. The disadvantage of working with radio waves is their long wavelength, which means that even a really huge radio telescope cannot match the resolution of a pair of binoculars. This problem is overcome by combining recordings from widely separated radio telescopes. In this way, the effective aperture (see *The Greenwich Guide to Astronomy in Action*) can be increased to thousands of kilometres, giving a very high resolution indeed.

The early surveys showed that there were several hundred radio sources scattered around the sky and attempts were made to correlate these with visual objects. This proved remarkably difficult because, apart from the Sun, the positions of radio sources did not coincide with bright stars. Deep sky photographs are crowded with fainter images, and the early radio positions were not good enough to decide which of these might be the source. However, when better radio positions became available a pattern began to

Centaurus A, the first radio source to be positively identified with an extragalactic nebula.

emerge. Many of the sources were identified with extragalactic nebulae—elliptical or spiral galaxies—many of which had streams of gas extending far beyond the spiral arms. It is not too surprising that some galaxies are radio sources— after all, the Milky Way is such a source—but what *is* surprising is how intense the signals are. Making due allowance for distance, it is found that Cygnus A (the first radio galaxy to be detected) is a million times more powerful as a radio transmitter than our Galaxy. High resolution observations show that most of the signal comes, not from the central nucleus, but from two zones equally spaced on either side of the nucleus and far outside the optical image. This double-lobe structure is quite common among radio galaxies, though some show only a central source and others have lobes as well as a central source. The reason for this is not clear, though it may have something to do with magnetic fields associated with the galaxies.

Some of the radio sources proved more difficult to identify, because all the images in their direction were starlike. By good fortune one of the strong but unidentified radio sources, designated

The radio emission from some sources, for example Centaurus A, comes from two lobes on either side of the visible image.

3C273 to indicate that it is number 273 in the third Cambridge catalogue, is close enough to the plane of the Moon's orbit for the Moon occasionally to pass in front of it. By carefully timing when the Moon cut the signal off and when it reappeared, and knowing the precise orbit of the Moon, it was possible to deduce an accurate position for 3C273. At this point in the sky there is a rather faint, bluish image which looks much like any other thirteenth magnitude star except that it appears to have a small jet sticking out of it. Another strong radio source, 3C48, is surprisingly close to a sixteenth magnitude star. These were at first thought to be radio stars as opposed to radio galaxies, but were cautiously called *quasi-stellar radio sources*, or *quasars* for short. They turned out to be very peculiar objects indeed.

The spectra of 3C273 and 3C48 showed bright lines that could not at first be identified with any known element. Then, in 1963, Maarten Schmidt realized that the pattern of the lines was the same as that in the ultra-violet spectrum of hydrogen, but greatly displaced towards the red end of the spectrum. If this red-shift was due to the doppler effect, it would imply that 3C273 was receding

from us at a staggering 48,000 kilometres per second and 3C48 at an even more incredible 110,000 kilometres per second—37 per cent of the speed of light. The only other objects known to have such speeds were very faint and remote galaxies, though at the time none of them could match the speed of the quasars. On the assumption that this speed is due to the expansion of the Universe, this made the quasars the most distant objects then known.

A thirteenth or sixteenth magnitude star is nothing out of the ordinary, but for the remotest objects in space to appear that bright, their intrinsic luminosity must be colossal. It is estimated that one quasar, 3C297, would be ten thousand times brighter than the Andromeda Nebula if it were at the same distance. But that is only a part of the problem. The brightness of quasars varies with time, typically over intervals of a week to a year. This puts a limit on their size, because the instruction to get brighter or to get fainter cannot be transmitted faster than the speed of light, so that an object with a radius of a light year or larger cannot show overall variations over an interval of less than a year. Thus quasars must have radii of much less than a light year. How on earth (or should I say 'in space') can such an object pour out ten thousand times as much energy as a galaxy with a radius more than ten thousand times larger? At present there is no satisfactory answer to this question.

To get round this difficulty, some scientists have suggested that the red-shift of quasars is not due to the expansion of the Universe. For example, they could be relatively nearby objects that have been shot out from our Galaxy at enormous speeds. This gets round the energy problem, but raises others. Where did the energy come from to accelerate them to these speeds? Why do we not see blue-shifted quasars that have been shot out of other galaxies in our direction? Alternatively, the red-shift may be due to something other than the doppler effect. This would allow the quasars to be nearby and slow

moving. One of the predictions of Einstein's theory of relativity is that the gravitational attraction of a dense star will produce a red-shift in the light it emits, but the star would need to be superdense to produce the red-shift observed in quasars, and the spectra of quasars are quite unlike those of dense stars. The consensus of opinion is that quasars really are what they appear to be—small, distant objects with an enormous power output.

The time taken for quasars to vary in intensity puts an upper limit on their size, but what about the radio sources that do not vary; are they effectively points, or do they have extended images? One approach was to see if they twinkled. It is well known that stars twinkle and planets do not. This is because small-scale irregularities in the atmosphere bend the light slightly; for a point source like a star this causes twinkling, but a relatively large image like a planet remains steady. The possible twinkling of radio sources was investigated by Antony Hewish in England, who had a radio telescope adapted to measure rapid changes in intensity. The results were quite unexpected. Jocelyn Bell Burnell, at that time a research student, was using the equipment in 1967 when she found that one of the radio sources was not showing the random changes in strength that would be expected from twinkling, but was transmitting a regular series of short sharp pulses, one every 1.34 seconds. Could this be some extra-terrestrial intelligent being trying to communicate with us? With this outside possibility in mind, the source was called an LGM, short for Little Green Man!

It was not long before similar sources were discovered; LGM2, 3 and 4, one with a period of only a quarter of a second. But the idea that they came from an extra-terrestrial intelligence did not last long. If they were signals from intelligent beings, they would be expected to come from a planet. The orbital motion of a planet within the Solar System would cause it to move against the star background, but no such movement was found. If the signals came from a planet going round a star other than the Sun, the orbital movement should have shown up as changes in the pulse rate, due to the doppler effect, but this did not happen either. The pulses were as regular as clockwork once our own orbital speed had been taken into account. So regular, in fact, that at one time it was thought that they might provide a more uniform measure of time than atomic clocks. The name LGM was dropped and such an object is now known as a *pulsar*, an abbreviation of *pulsating radio source*.

As more radio observatories obtained suitable equipment, many more pulsars were discovered, with periods ranging from a few hundredths of a second up to four seconds. Most of them are in the plane of the Milky Way, and this is strong evidence that they are within our Galaxy. Although regularly spaced, the pulses are not all equally strong. Sometimes they disappear altogether, but when they return they are exactly 'on beat', as though they had never left off. If you look at them for long enough, however, you will see that the period of the beats is not quite constant. Careful measurements show that all pulsars are very gradually slowing down; that is to say, the intervals between pulses are slowly increasing. Some also show aberrations known as *glitches*; their rate of pulsing suddenly speeds up and then resumes its steady slowing down at the same rate as before the glitch. It was some time before any pulsar could be identified with a visual object, but eventually the pulsar in the direction of the Crab Nebula was identified. This is the fastest pulsar of all, sending out thirty pulses per second. But what exactly are pulsars? To answer the question

This delicate tracery of gases in the constellation Vela is the remnant of a supernova that exploded some 10,000 years ago. The Vela pulsar (not visible here) is the stellar remnant of the supernova. It pulses twelve times a second and provides the radiation which causes the gas to glow.

it is necessary to know in more detail what happens to a star when it gets old.

As we have seen, a star like the Sun, over a period of thousands of millions of years, will eventually run out of fuel and start to cool into a white dwarf. It is the heat generated within a star by nuclear fusion that keeps it inflated. Once the central furnace gets turned off, the star will start to shrink. The shrinking process itself provides a source of energy with gravitational energy being converted into heat energy as the star collapses. Thus, while the star is getting smaller it is still white hot, but this cannot go on for ever. The energy is being radiated away into space in the form of heat and light, and it is not being replaced, so the star will just go on shrinking until it can shrink no more, and then cool down to the temperature of space, just a few degrees above absolute zero.

At this point it will be something like a giant ball-bearing, but that is not the end of the story. A ball-bearing with the same mass as the Sun would have a radius about 20 per cent bigger than that of Jupiter and, because so much mass is concentrated into such a small volume, the gravity at its surface will be two thousand times greater than on Earth. A man there would weigh well over 100 tonnes. This may seem bad enough, but conditions are worse inside. The inner layers are required to support the outer parts of the ball-bearing against the enormous downward pressure of gravity, but they are unable to do so. The whole ball, even though made of iron, would compress itself into an even smaller ball.

The reason you cannot normally compress a lump of iron is because of the forces between its

The constellation of Orion dominates the northern hemisphere winter sky. It is about the size of a postcard held at arm's length. The red giant star Betelgeuse (top left) contrasts with the blue-white star Rigel in the opposite corner. The Great Orion Nebula is the red patch in the sword of Orion, hanging below the three stars that form his belt.

atoms. Each atom consists of a small, central nucleus (which contains most of the mass) surrounded by a cloud of electrons. The electrons effectively form a barrier round each nucleus, so no two nuclei can get closer than the barriers will permit. In our massive ball-bearing, the gravitational forces are more than the barriers can withstand with the usual distribution of electrons, so the electrons re-group to form a smaller but stronger barrier which is able to stop the collapse. They are then in a state known as *electron degeneracy*.

A white dwarf ends up as a ball of degenerate matter about the size of the Earth, but so dense that a single teaspoonful of it would weigh 5 tonnes. It would take a strong man to lift up a pin made of this material.

Even degenerate material does not have infinitely strong barriers, as its resistance to compression depends on the orbital speed of the electrons as well as their distribution. Even if the electrons were to approach the speed of light, they would not be able to withstand the gravitational pressure from a shrinking star with a mass of more than 1.4 times that of the Sun. As we saw when we considered planetary nebulae, some stars that start off with too much mass are nevertheless able to end up as white dwarfs by exploding off their outer layers to form an ever-expanding shell and leaving behind a core that is less than 1.4 solar masses. This will work with stars that are not too massive to start off with, but what about the even more massive ones?

A star of about 10 solar masses goes through many of the same processes as a smaller one, though rather more ostentatiously. It starts off on the main sequence, though higher up than the Sun because it is brighter and hotter. There it burns up the hydrogen in its core in double-quick time (mere tens of millions of years rather than thousands of millions) and evolves off the main sequence to become a *red supergiant*. The best-known naked-eye example is Betelgeuse, the red star in the armpit of Orion, the Hunter. While

shell-burning of hydrogen is taking place, the core itself rises to higher temperatures and produces further energy by fusing helium into still heavier elements.

When helium is formed from hydrogen, energy is given off. This is also the case when helium is fused into heavier elements, but there will only be energy output if the product element has less mass than those from which it is formed. At the very heavy end of the scale the process is reversed. When uranium is broken up into two lighter elements, their combined mass is less than that of the uranium atom from which they come, so energy is released. This is the principle of the atomic bomb. Conversely, energy would be needed to form a uranium atom out of two lighter ones. Iron lies in the middle of the scale. This metal *can* be split into lighter elements, or combined with something else to make heavier elements, but either process will result in a net loss of energy.

The chain of conversion from helium into heavier elements runs through carbon, oxygen, silicon and calcium to iron, at which point the core of a supergiant will no longer be a source of energy, but a drain. The core will then start to shrink under its own gravity and, as it does so, it will get very hot. Some of this heat energy will be used up in converting the iron into other elements and this will only accelerate the collapse. Things then happen very rapidly by any standards. In a matter of seconds, the core will collapse and become unimaginably hot. Material from further out in the star will fall towards this furnace and its subatomic particles (protons and electrons) will be welded together to produce neutrons and other subatomic particles called *neutrinos*. Neutrinos are curious particles that have almost no mass at all and travel at very nearly the speed of light. They are very difficult to detect because many millions of them can pass straight through a simple obstacle like the Earth. Only one or two out of many millions will get stopped and these can only be detected with the most sensitive of apparatus. But the layers enveloping the star's collapsing core are so dense that they are able, very briefly, to interrupt the outward flow of about one per cent of the neutrinos. In doing so, these layers receive an enormous outward thrust, so great that the whole star is blown to pieces in an explosion second only to the Big Bang—the star becomes a *supernova*.

Until recently, this picture of the mechanism leading to a supernova was rather speculative. Although it is estimated that on average one supernova goes off every second somewhere in the Universe, the last one in our Galaxy was nearly 400 years ago, before Galileo had built his first telescope. Astronomers were desperately keen for one to occur nearby so that they could test their theories. On 23 February 1987 their wish was granted. A supernova exploded in the Large Magellanic Cloud. In many ways this is even better than having one in our own Galaxy, because it is close enough for detailed observation and yet is not obscured by the dust and gas in our galactic disc. Measurements are still being made and interpreted and there are many features that are not yet understood, but already it is clear that the theory is not too far from the truth. In particular, a burst of neutrinos was detected at the time the supernova exploded, confirming the most controversial part of the mechanism. One surprise is that the progenitor star was blue rather than the expected red, but it is not too hard to reconcile this with the theory and probably explains why the light curve differed from those observed for supernovae in more remote galaxies. The 1987 supernova brightened very rapidly and then remained steady for some time, whereas a red supergiant would have risen more slowly to a maximum and then started to fade.

Besides being spectacular, supernovae are of

Supernova 1987a in the Large Magellanic Cloud. The lower picture shows the same star field before the supernova explosion.

great importance for astronomy. In one second they release a hundred times more energy than the Sun will in its entire lifetime, and it is in that second that all the heavy elements are formed, to be dispersed through the galaxy by the explosion. Also associated with the explosion are violent shock-waves, somewhat like the sonic booms from supersonic aircraft (though on a titanic scale) and these are believed to be the trigger that starts the condensation of stars out of gas clouds. It is safe to say that without supernovae we would not be here.

At its brightest a supernova outshines all the stars in its galaxy, but even after it fades it is still a spectacular object. The best known supernova remnant is the Crab Nebula, probably the remains of an explosion that occurred in 1054, when the Chinese recorded a 'guest star' which was bright enough to be seen in daylight. Now it is a tortuous mass of gas filaments, over 4 l.y. across and still expanding. It is also a source of

Computer-enhanced, false-colour image of the expanding shell of material from Tycho's supernova.

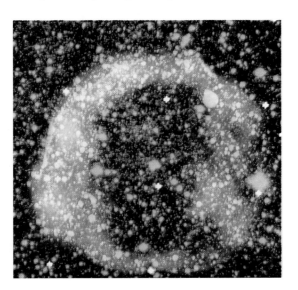

radio and X-rays. Other supernovae occurred in our Galaxy in 1572 and 1604, known as Tycho's supernova (after Tycho Brahe) and Kepler's supernova (after Johannes Kepler) respectively.

But what becomes of the core of a supernova? It is now composed only of neutrons, with no circling electrons to keep them apart from one another. The core has turned into a *neutron star*, which can become much smaller and denser than even the degenerate material of a white dwarf. Whereas a white dwarf with the mass of the Sun would be about the size of the Earth, a neutron star of the same mass would be only 15 kilometres across—about the size of a large city—but a teaspoonful would weigh nearly a thousand million tonnes.

As the core of the supernova contracts, two things are conserved: its spin and its magnetic field. Neither of these need have been anything out of the ordinary before the collapse, but even a small amount of spin gets greatly amplified when it is compressed into a small body, just as an ice skater who starts to spin with his arms extended sideways will then spin faster if he raises his arms above his head to make himself narrower. Most stars have at least a small magnetic field and this is affected in the same way, so we would expect a neutron star to spin very fast indeed and also to be a strong magnet. Curiously, a neutron star does not have to be more massive than a white dwarf. Since all its electrons and protons have been transformed into neutrons, even a small neutron star can collapse without experiencing any barriers other than those provided by movements due to temperature, and it is only a matter of time before it cools down and becomes tiny.

The mysterious pulsars are in fact these neutron stars. The pulses of energy come, not from the star itself, but from electrons trapped by

The spiral galaxy NGC 5236, also known as the Southern Pinwheel. No fewer than four supernovae have been seen in this galaxy since 1923.

its magnetic field (there are still plenty of electrons outside the star), which produce what is known as *synchrotron radiation*. When an electron is left alone, it travels in a straight line, but if it encounters a magnetic field it will be deflected. If you put a magnet up against a television screen, it will distort the picture by deflecting the electrons which make the screen glow. If an electron is travelling very fast and is strongly deflected, it will emit light, but not uniformly in all directions. The light shines in the direction the electron would have travelled had the magnetic field not been there. Actually, it is a cone of light, whose angular width depends on the speed of the electrons—the faster they are moving, the narrower the cone. Synchrotron radiation can be in the form of X-rays or radio waves as well as visible light, and all these have been detected from pulsars. The signals are intermittent because the light-cone rotates with the star, rather like the light from a lighthouse which we see only briefly as the beam sweeps

round. We then have to wait until the beam completes a revolution before we see it again. And we have to be within the area of the cone in order to receive the signal. There are probably very many pulsars that we never see because their cones miss us completely.

The picture now begins to fall into place. The pulsar in the direction of the Crab Nebula is the original core of the supernova of 1054 and it is spinning very fast (thirty times a second) because it is still young. As it slows down, the intervals between pulses will increase. If the neutron star rearranges its shape to become still more compact, it will speed up (like the skater) and we will see a glitch.

The other pulsars are similar neutron stars formed in other supernovae. They have to be neutron stars because a bigger object, with a lower surface gravity, would not be able to spin so fast without breaking up. For example, if we could increase the rate of spin of the Earth until a day was only 90 minutes long, the Equator would be moving at the speed of low-orbit satellites, at about 7.5 kilometres per second. We could launch a satellite from the Equator by giving it a gentle push, but we would have to take care not to jump or we would go into orbit ourselves. If we tried to speed the Earth up even more, anything near the Equator—people, buildings, mountains or ocean—would fly off into orbit. At one revolution every second, the whole Earth would break up and be flung into space.

We have seen that neutron stars can have quite a low mass, but is there an upper limit? Neutrons are not infinitely strong and even they eventually get squashed if the supernova core remnant is greater than about three solar masses. In this event, very strange things happen. Neutrons present the final barrier. When this is breached, we know of no force to prevent

The Crab Nebula. This rapidly-expanding shell of gas and the pulsar within it are the remains of the supernova of 1054.

gravitational collapse continuing until the star is all concentrated at a single point, or *singularity* as it is called. The gravitational attraction of any body (the Earth, for example) will increase as you get close to its surface, but it diminishes from the surface towards the middle. With a *black hole* (for this is what we are talking about) there is no surface, so the gravitational attraction just goes on increasing towards infinity as you get closer. Anything heading directly towards a black hole would get sucked into the singularity. However, if you were in a spacecraft, there is no reason why you should not go into orbit around a black hole, providing you take care to choose an orbit that does not get too close.

What you would see would be determined by the effect gravity has on light. This was first worked out by Albert Einstein in his general theory of relativity, and has since been tested experimentally, so we have some confidence in the results. In essence, there are two effects. As we saw when looking at pulsars, the gravitational field of a star will red-shift the light that is leaving it. The other effect (which has not been mentioned before) is that light passing close to a massive star will be bent towards it. For a normal star like the Sun, the red-shift and deflection are so small that they are difficult to detect at all, but a black hole is not a normal star.

Imagine that there is a laser on your spacecraft and that there is enough dust around to show up the path of its beam. As the laser beam is swung round towards the black hole, it will be bent towards it. The closer it gets, the more the deflection until, if we aim carefully enough, it will be bent round the back of the black hole and shine back at us. If we go very slightly further in, say to a distance of 45 kilometres from a black hole ten times the mass of the Sun, the beam will be bent right round in a circle and will never leave the black hole. Any closer than this, and the beam is just swallowed. The sphere around a black hole that is able to capture light in this way is called the *photon sphere*.

What happens when we drop a flare into the black hole? We will have to have some means of recording the result because the fierce gravity will suck the flare through the critical last few tens of kilometres much faster than the twinkling of an eye, and the red-shift will be so great that the light will be well outside the limited range that our eyes can detect. As the flare approaches the black hole it will get redder and redder, but it will still be visible even after it passes through the photon sphere. But once the flare gets closer to the centre than two-thirds of the radius of the photon sphere, no light will escape at all. It will have passed through the *event horizon* which, for a black hole of ten solar masses, is 30 kilometres in radius. At this distance from the centre, the escape velocity is equal to the velocity of light, so absolutely nothing can ever get out: no light, no radio waves, no particles, no information of any kind. That is why it is called a black hole.

Black holes and their properties were originally predicted theoretically, but how do you set about finding a real one in the sky? This might seem like an impossible problem, since nothing ever comes out, but the gravitational attraction of a black hole should have an effect on nearby material. For example, if a black hole is one component of a double star, it might be possible to detect its presence by seeing its companion apparently orbiting around nothing. Alternatively, if the black hole is near a cloud of gas, it might be expected to drag some into orbit around it and accelerate the gas to such high speeds that it will give off X-ray radiation. Several examples have been found that show both these effects.

The most likely candidate is the X-ray source known as Cygnus X-1: Cygnus because that is the

Artist's impression of Cygnus X-1, a system which is thought to include a black hole and a blue supergiant star. The intense gravity of the black hole draws gas from the supergiant companion to form a disc of high-temperature plasma around the black hole.

constellation in which it appears and X-1 because it is the first X-ray source to be found in that constellation. The X-rays were detected by the UHURU satellite launched by NASA in 1970, and have been found to vary in intensity over intervals of a few thousandths of a second—a promising start. Cygnus X-1 is also a radio source. In the same position in the sky there is a ninth magnitude blue supergiant with a mass about fifteen times that of the Sun. Its radial velocity has been measured from a series of spectra, and is found to vary with a period of 5.6 days. This is exciting because it shows that the star is a binary, although no companion is visible. As we saw in Chapter 3, it is not possible to deduce the mass of the companion from radial velocities alone, but the most likely estimate is about eight solar masses, with four as the lower limit (unless there is a third star in the system). This is much too massive for a neutron star. All these results strongly suggest that Cygnus X-1 contains a black hole, but, tantalizingly, we can still not be certain.

There are two other likely black holes in the Large Magellanic Cloud—LMC X-1 and LMC X-3. Like Cygnus X-1, these are both X-ray sources and both involve a blue supergiant orbiting a compact but massive companion. Then there is the transient X-ray source called A0620-00, which flared up briefly in 1975 to become the brightest celestial X-ray source ever recorded.

This has recently been found to be a low-mass, red spectroscopic binary. The mass of its companion is at least 3.2 solar masses. This candidate is particularly interesting because, unlike Cygnus X-1, the estimated mass of the companion cannot be reduced by postulating the presence of a third star in the system.

We have seen how stars might give birth to black holes of a few solar masses, but it is also possible that massive black holes, equivalent to millions of stars, might be present at the centre of galaxies and be the source of the X-rays and radio emission that are seen to come from them. At present there are two strong candidates. One is the dwarf elliptical galaxy M32 which is close to the Andromeda Nebula. The other is our own Galaxy (see Chapter 5). Recent radio observations have shown that the source at the centre of our Galaxy has a diameter no greater than about twenty astronomical units—the size of the Solar System out to the radius of Saturn.

Once a black hole has formed it is likely to grow, swallowing up anything that strays into it. As it grows in mass its event horizon expands, though the mass within it is still confined to a point. Black holes may be very common, but they are so difficult to detect that we are unaware of them. Indeed, it may be the fate of the Universe to collapse back on itself in a sort of inverted Big Bang to end up as one super-massive black hole, though still smaller than a pinpoint.

7 · Why Bother?

On his return from leading the first successful expedition to climb Mt Everest, Sir John Hunt was asked why he had done it. His reply was 'because it is there'. This might equally well be the answer to the question 'why study astronomy?', but in fact there are many more substantial reasons for doing so.

The weakest argument would be that the study of astronomy has yielded many useful results. For example, but for astronomy, we would know very little of how gravity works and so would not have been able to launch the artificial satellites that give cheap, worldwide communication, safe navigation, weather, fisheries, crop and iceberg information, as well as many other advantages. As an example of benefits yet to come, it was through astronomy that hydrogen fusion was discovered and this process is the most promising source of clean, safe energy for when fossil fuels run out. But this is similar to using non-stick frying-pans to justify the American space project. Substantial spin-offs are certainly there, but it is not for these we study astronomy.

A scientist will recognize that the heavens provide a laboratory where he can study processes under conditions that can never be matched on Earth. No laboratory vacuum can approach the emptiness of space, or any high pressure in the laboratory come anywhere near that found in white dwarfs, let alone neutron stars. Low temperature physicists can get nearer to absolute zero than the temperature of space, but where high temperatures are concerned we cannot come anywhere near those generated by stars. The main disadvantage of the heavenly laboratory, and it is a big one, is that we can only be passive observers of experiments that we did not design. Against this drawback can be set the huge number and diversity of the experiments that are going on. If one star does not provide the exact conditions we require, we just look around until we find one that does.

But astronomy is not simply an extension to an earthly laboratory; it is well worth studying in its own right purely for the intellectual challenge it provides. It is like the most fiendishly complicated crossword puzzle ever devised, with only the most subtle of clues. Frequently there is more than one possible answer and we cannot be sure which is right until we look at the evidence from another direction. Sometimes we fill in an answer and later have to rub it out because it does not fit in with something else. Also like a crossword, the enjoyment is in searching for the answers rather than in finding them. Once we have solved a clue we may allow ourselves a few moments satisfaction at how elegantly the answer matches the clue and how well it fits in with the other answers; then we fill it in and move on to the next problem. Unlike a crossword, we are never likely to complete the astronomical puzzle. Some sections are complete and others very sparsely filled in, but we can never be sure that we will not have to go back for a complete re-think when we find ourselves left with a five-letter word containing two Qs and one Z.

Much of the pleasure of astronomy, then, lies in the 'thrill of the chase'. Astronomers are never happier than when they have two different theories that both fit the existing evidence and

further observations of a different kind are needed to distinguish between them. For example, in the 1950s the expansion of the Universe and many other large-scale phenomena could equally well be accounted for by the Big Bang theory or by the *steady state theory.*

The steady state theory was proposed by Fred Hoyle, Hermann Bondi and Thomas Gold. They suggested that the Universe had always looked much as it does now, and always will. Since it is expanding, new galaxies must come from somewhere to fill in the gaps, otherwise the density of the Universe would decrease and would not be in a steady state. Their brilliant and revolutionary idea was that creation did not occur in a single event, but was and is a continuous process. A new particle might spring into existence from nowhere, anywhere and at any time in the Universe, and the *continuous creation* of matter takes place at exactly the right rate to counterbalance the loss of density due to expansion. (Alternatively, the expansion is at just the rate required to avoid an increase in density.) The new material would then condense to form galaxies and stars, and the process could go on for ever. This elegant theory was worked out in considerable detail.

Both the Big Bang and steady state theories had their supporters, though perhaps the majority of astronomers preferred the latter, if only because it removed the need for a singularity, never an easy concept to handle. The first observational evidence that could distinguish between the rival theories was produced by the radio astronomer Martin Ryle. In looking at far distant objects like radio galaxies, you are also looking backwards in time, because of the time it takes the signal to get here. By taking account of the rate of expansion of the Universe, Ryle inferred that the density of radio galaxies had been greater in the past than it is now, which would be expected for a Big Bang but not for a steady state Universe. More conclusive evidence came later, when Arno Penzias and Robert Wilson detected the microwave background radiation which could be explained only by the Big Bang theory.

So the steady state theory had to be abandoned (except by a few diehards) and the science of cosmology had taken a major step forward. But most people would agree that those were stimulating times and that much of the excitement went out of cosmology when the controversy was resolved.

In attempting to compress much of the Universe into a small book, I have had to avoid including too many 'buts' and 'maybes'. By omitting many of the alternative theories and stating the currently fashionable ones as though they were firmly established, I may have done the subject a disservice by making it seem more cut-and-dried than it really is. Every branch of astronomy has a base of reliable information surrounded by a frontier at which we are hacking away. And beyond the frontier, the great jungle of the unknown. Even the home-base may not be as secure as we think. Nothing could have been more confidently believed than Newton's theory of universal gravitation until Einstein came along and showed it to be merely an approximation, though admittedly a very good one.

So astronomy is an intellectual challenge, but it is more than that because it is about nature itself. Ultimately with its aid we can hope to address some of the most profound questions of all, about where we come from and where we are going to, or, as *The Hitch-hiker's Guide to the Galaxy* put it, the question of life, the Universe and everything. We must be careful not to stray out of science into the realms of religion. Astronomy can never answer questions about the purpose of the Universe, but it can give us great insight into how the Universe works. It has been argued that we can never comprehend the Universe as a whole because its complexity is more than can possibly be accommodated in the very limited capacity of a human brain. Put in mechanical terms, can a machine be devised to

make other machines that are more complicated than itself? If there is such a limit, we have a long way to go before we reach it and even if there are some questions that can never be answered, it is still extremely stimulating to try.

Astronomy is not only about mind-stretching, though. Most astronomers, professional as well as amateur, find it both relaxing and rewarding simply to enjoy the glorious spectacle provided by the heavens, much as they might enjoy listening to music. Just as the pleasure of music can be enhanced by knowing something of its background, so it helps to have some idea of what is going on in the sky. It is hoped that the 'programme notes' provided by the *Greenwich Guides* will add to your enjoyment of the night sky, but they should not be used as a substitute for going outside and looking at the real thing.

Index